Trench excavation and support

Trench excavation and support

J. K. Budleigh

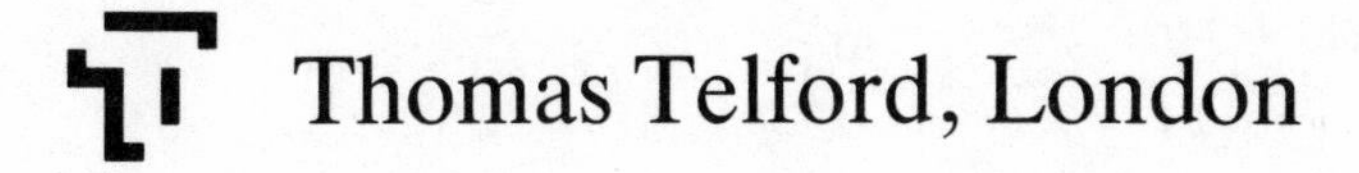

Published by Thomas Telford Ltd, Thomas Telford House, 1 Heron Quay, London E14 9XF

First published 1989

British Library Cataloguing in Publication Data
Budleigh, J. K.
Trench excavation and support.
1. Trenches. Construction
I. Title
624.1'52

ISBN: 0 7277 1347 7

Typeset in Great Britain by Pentacor Ltd, High Wycombe, Bucks
Printed and bound in Great Britain by Billing and Sons Ltd, Worcester

Contents

1. Introduction 1

2. Route selection and early planning 5
Longitudinal profile; land interests and compensation; physical destructions; areas of bad ground; future access needs; existing public utilities and structures; social costs

3. Site investigation 10

4. Choice of excavation plant 13
Continuous trenching machines; jib operated excavators; applications

5. The causes of trench failure 19

6. Traditional methods of trench support 26
Support by poling boards, waling and struts; design of traditional trench supports; support by interlocking sheet piles; soldier piles; staged trench excavation

7. Proprietary support systems 38
Hydraulically strutted shores and walings; shields and boxes; plate lining systems

8. Dealing with groundwater 43
Sump pumping; well pointing

9. Bottoming-up and backfilling 48

10. Reinstatement 53
Highway reinstatement; reinstatement of unpaved land; reinstatement of land drainage; reinstatement of individual drain pipes; construction of an interceptor drain

11. Contractual aspects and safety 61
The contractual relationship; safety hazards in trench, excavation; the human element; compliance with construction regulations

Acknowledgement 72

References 73

1 Introduction

Trenches have been excavated for many purposes, not all connected with civil engineering. In the normal engineering sense, however, their purpose is to allow installation of public utilities and sewers and this book has been written with these applications predominantly in mind. The apparatus most often installed is a pipeline of some kind and, for simplicity, it will be assumed that this is the case.

As an indication of work volume, the total length of public utility main pipelines in the UK is approximately 850 000 km without counting commercial pipelines and trenches excavated for electricity and telecommunication cables. These existing trenches represent an enormous investment — over £100 000 million in the UK at 1988 prices — hence the concern over past neglect of underground assets and the pressure for cheaper innovative renovation methods.

Because they are laid to serve populated areas the majority of utility services are in public highways. However, some of the most important trenching projects are on cross-country routes or undertaken in exceptional and difficult environments such as sea beds. A few specialist techniques will be touched on, but this book will be concerned primarily with the excavation methods most commonly in use and with trenches up to 6 m deep.

It is normal for trenches to be dug to as narrow a profile as possible, consistent with placing and backfilling around the installed apparatus. Therefore trenches are normally cut with vertical sides. However, in cross-country routes where the trench is relatively deep and reinstatement costs cheap, a battered profile may be economical (Figs 1 and 2).

As depth increases there comes a point at which tunnelling begins to compete in cost with trench excavation and the case for tunnelling is further enhanced in built-up areas if account is taken of the social costs of the disruption caused by open trench work.

Tunnelling is nowadays a broad generic term which covers a great variety of construction methods. One of those most directly competitive with open trenches is pipe jacking, which is a well proven system in which a robust, thick walled pipe, normally of reinforced concrete, is thrust into the earth face. The earth face is excavated manually, at the same time, from inside the pipe (Fig. 3).

Pipe jacking is often suitable for sewer construction at depths greater than 6 m, and it can also be used to provide ducts for pressure pipelines beneath special obstructions such as railways and motorway embankments.

In addition to pipe jacking, other forms of small bore tunnelling have been developed which, together with moling and no-dig renovation techniques, have the effect of shifting the point of economic balance further away from open trench construction. This is particularly so where the aim of the project is replacement or renovation of an existing delapidated pipeline. However,

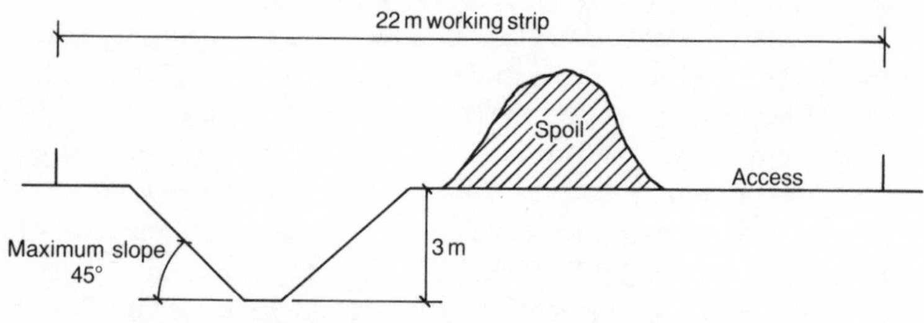

Fig. 1. Battered trench

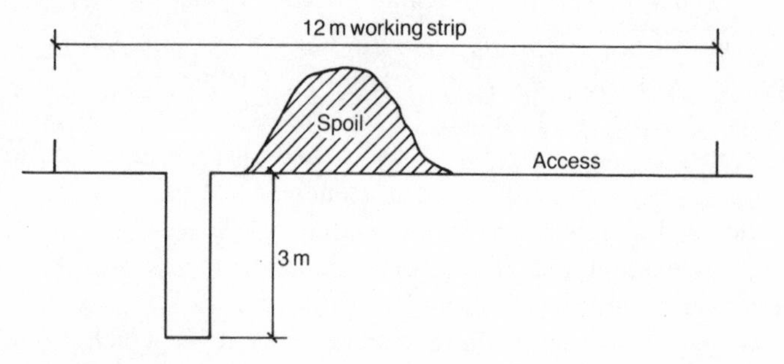

Fig. 2. Vertical side trench

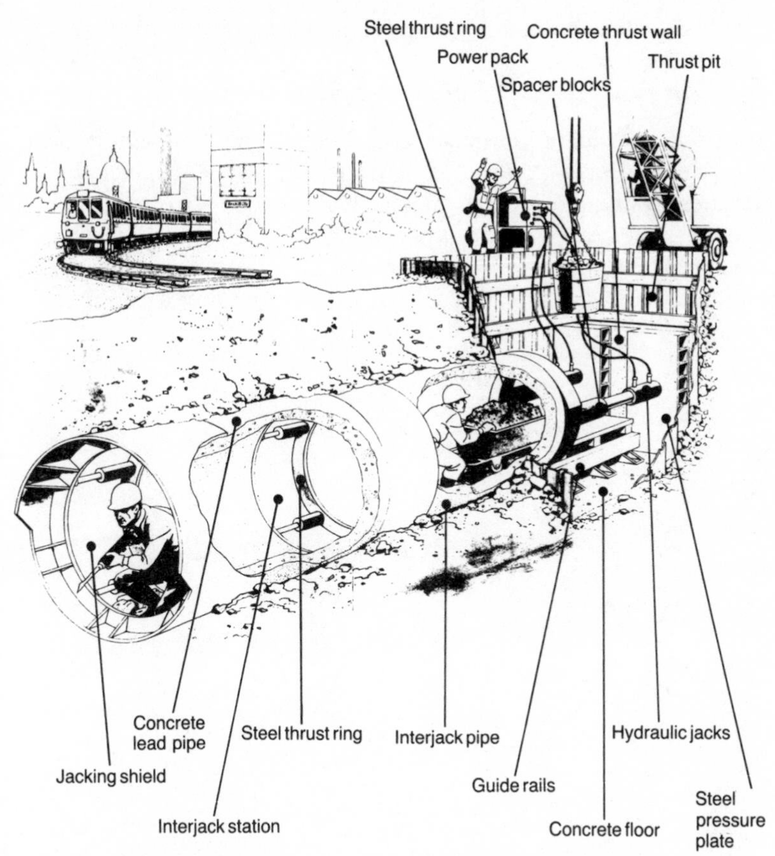

Fig. 3. Pipe jacking

despite this trend open trench excavation remains the most economic method for installing most sewers and utility services laid at normal depths and will continue to represent a regular and important part of the civil engineering industry.

The excavation of the trench is, of course, a very simple matter whether by hand or machine. The skill lies in excavating in such a way that

(*a*) the trench sides are adequately supported
(*b*) groundwater does not interfere with the work
(*c*) the bottom surface is prepared and the apparatus placed and backfilled in such a way that both the installed plant and

any adjacent buildings and services are safe against further movement

(*d*) the ground surface is properly reinstated.

Above all the operation must be carried out safely. Regrettably this is not always achieved and trenching accidents, including fatalities, continue to feature regularly in the published statistics. Safe working methods are not new and the technology involved is comparatively simple. What is required to avoid further tragedies is the constant vigilance of supervisors and operatives and more consistent application of the best working practices.

2 Route selection and early planning

Unlike most civil engineering projects, trenching schemes are not confined within a recognized site but, by their nature, lie between widely spread points of connection and involve working in a variety of rural and urban enviroments, often in close contact with the public and with third party land interests. One of the first tasks for the engineer is to select a suitable route.

At the smaller end of the market, distribution pipelines serving homes, which account for the majority of all trenches, require very little planning as the route is largely predetermined by the layout of properties.

For larger pipelines the many factors which influence the route can be grouped under the headings

(*a*) longitudinal profile
(*b*) land interests and compensation
(*c*) physical obstruction
(*d*) areas of bad ground
(*e*) future access needs
(*f*) existing public utilities and structures
(*g*) social costs.

Longitudinal profile

The importance of longitudinal profile will depend on the type of pipeline. For gravity sewers it is critical as the sewer gradient will rapidly result in deep excavation and pumping unless the best advantage is taken of existing ground slopes. More freedom is possible with pressure pipelines but the profile may be constrained by the need to keep the pipe level below the hydraulic gradient for all operating conditions and perhaps, in hilly areas, by the need to avoid excessive pressure in the lower levels. Very steep sections may pose problems in construction and long-term stability. A study of contoured Ordnance Survey maps at an early stage of planning will help to avoid such difficulties.

Land interests and compensation

Promoting authorities normally have strong statutory powers enabling them to plan pipeline routes on engineering and economic grounds without being unreasonably deflected by hostile land owners. These powers do not apply to railway or canal property or to Crown land where entry can only be gained by consent. For commercial pipelines *ad hoc* statutory powers with similar force may be acquired by special Act of Parliament. The legislation also provides for compensation to owners and occupiers for surface damage, which is normally a fairly modest sum, and for damage to the freehold caused by the presence of the pipeline which is roughly proportional to land values. Thus routes through land with the prospect of development or of mineral extraction will be expensive and should be avoided where possible. The advice of a professional valuer should be taken where uncertainty exists.

Routes through woodland require particular care not only because of tree felling costs but because the permanent sterilized strip along the pipeline route can seldom be replanted and may become a significant maintenance liability.

Apart from considerations of compensation cost it should also be recognized that pipelines carrying certain materials, for example, high-pressure gas are perceived by the public as a potential danger, which indeed they are. Such pipelines should be kept well clear of residential property.

Physical obstructions

Obstacles such as existing railways, canals and motorways can influence the route significantly as there may only be a limited number of feasible crossing points. It is important that negotiations with the responsible authorities are opened as early as possible in the planning process to avoid a loss of valuable time and the possibility of late route changes.

Areas of bad ground

Areas of bad ground may be found by inspection of Ordnance Survey and geological maps, aerial photography, in the case of larger schemes, and by local inspection and enquiry. Points to watch for include cavernous ground in limestone or chalk, steep slopes, evidence of land slips or other surface instability such as movement of established trees, spring lines or a high water table coupled with granular or sandy ground material, hill peat, lowland bogs and any recognizable geological fault.

Areas of mining subsidence require special consideration and can normally be discovered from the mine operators or from mine abandonment plans. In cases of large mine workings subsidence may have to be allowed for in the pipeline design rather than avoided altogether.

Future access needs

On cross-country routes an essential requirement is long-term access to the line both for repair and for working esssential equipment such as valves. This is an operational matter requiring consideration of return frequencies, alternative feeds, emergency response times and mobility of repair plant. The route should take account of these needs.

Existing public utilities and structures

For trenches in public highways the main problem is avoiding existing underground pipes and services. The machinery for this is the Public Utilities Street Works Act 1950 which provides for service of notices and negotiations between the utilities and between them and the highway authority. Normally the presence of apparatus is reasonably well known but the utility cannot guarantee its precise location. Where services are congested a detailed survey is necessary. In the case of sewers the line and level can normally be established quite accurately by lifting manhole covers but other services present more difficulty. Valve surface boxes can give clues and some pipes can be found by electro-location equipment. In critical areas, however, the only way to be certain is to excavate a number of trial holes, normally by hand, across the proposed trench route. There are practical limits to the frequency of these excavations but they should be sufficient to satisfy the design engineer that the trench can be excavated without undue conflict with existing services. In exceptional cases it may be necessary to move one or other of these services to make room for the proposed works. This is normally possible but naturally the promoter is expected to bear the cost.

An important consideration in urban areas is the effect of the proposed excavation on the stability of adjacent services and on the foundation of adjacent buildings.[1] Even where the best construction practice is followed in the support of the trench sides it is difficult to prevent some movement of adjoining ground. In many cases this may not be serious, but in others it can be critical, so the risk should be reduced as much as possible by leaving the maximum possible space between the trench and the installation at risk.

Similarly, account should be taken of the best distribution from the foundations of existing buildings and as a general guide trench sides should be kept clear of a 45° projection from foundation bases.

Social costs

A unique aspect of trench excavation in urban areas is the impact of the works on the community — what has become known as the social cost of the project.[2] This occurs in two forms: first in disruption and delays to traffic, the total cost of which in terms of lost time and vehicle running costs can be surprisingly high, frequently higher than the cost of the project itself. Second, in loss of business to commercial premises along the route: this again can be significant and, unlike traffic cost, will be concentrated on a limited number of businesses. The promoter pays a public relations penalty on both counts but seldom has a direct legal liability to those affected. However, where business losses are incurred the moral obligation is such that *ex gratia* payments are often made. Route planners should take a broad view of their responsibilities and recognize these social consequences even if they do not directly affect the project cost.

These then are the factors to be considered in deciding the route of a trenching project. The amount of effort put into planning depends on the type and magnitude of the project. In some cases the route is obvious from the start, but for most large cross-country schemes a detailed desk study is needed, followed by field surveys. Often several routes are feasible and must be worked through and costed before a choice is made. At this stage discretion is needed in dealing with land owners. It is pointless to alarm them unnecessarily on non-preferred routes; on the other hand it is wrong to be over-secretive. Commonsense is the most important ingredient in this and all other aspects of route planning.

When the route has been decided, formal contact must be made with owners and occupiers of land affected and with the authorities responsible for highways and public utilities. In most cases powers to execute the project will be given by legislation and appropriate statutory notices must be served.

The legislation also gives the owners rights to compensation for damage caused, so it is important to record in some detail the condition of the land before excavation. This record should be agreed at the time with the owner or occupier. In the case of work in public highways a record should also be made of the condition of surfaces to be excavated noting, for example, the proportion of footpath slabs already cracked. The condition of any structures or boundary walls close to the trench route should be recorded and preferably photographed at this stage.

3 Site investigation

A preliminary form of site investigation was mentioned in chapter 2 in the location and avoidance of bad ground when planning the pipeline route. In the normal sense, however, site investigation means the exploratory work, in the form of borings, trial excavation and soil tests done during contract preparation to determine how the ground is likely to behave when excavated.

Site investigation is the subject of an ICE Works Construction Guide[3] and will be considered here only in its particular relation to trenching projects. For these schemes the main purpose is to supply the contractor with basic data on which decisions on temporary trench support, choice of plant, and on the need for pumping and dewatering can be based.[4,5] In these respects there is no difference from most other civil engineering schemes. Unlike most projects, however, the bearing strength of the soil is unlikely to be critical in a trenching scheme unless the soil is exceptionally weak. On the other hand the engineer will have an interest in two other qualities of the ground which may not apply elsewhere. The first is the suitability of the excavated material as backfill to installed apparatus. Friable or granular material is generally suitable but large lumps of stiff clay, rocks or sharp stones are not. The second is the future action of the backfill on the material of the pipe. Stiff clays with low pH are unsuitable against iron or steel pipelines without adequate protection, and sulphates can attack concrete or asbestos/cement products.

A frequent subject for debate, particularly after trouble has occurred, is the amount of effort which shall be put into site investigation and there are particular problems here for the engineer on trenching schemes. The majority of these are of such a length that practical limits have to be set to the frequency of test samples. Also trench excavation is often shallow, less than 1 m, and it may then be sufficient to excavate trial pits at, say, 500 m centres rather than use full-scale site investigation methods. Indeed, in areas where the ground is known to be homogeneous and the engineer has sufficient local experience, even these trial excavations can be dispensed with.

On the other hand where trenches are deep — more than about 1·5 m — and/or where ground conditions are suspect or high water levels are known to exist, sufficient borings should be taken to supply the engineer and the contractor with the information they need.

The main factors to be determined by the site investigation are

(*a*) the strength of the soil
(*b*) the groundwater level.

The method of measuring the soil strength varies according to the soil type, i.e. depending on whether or not it is possible to obtain undisturbed soil samples, but in either case it is the shear strength of the soil which is measured.[6]

Where undisturbed samples are possible, in cohesive soils such as clays, the cylindrical soil sample is tested in the laboratory by compression to failure and its undrained shear strength (*Cu*) measured.

In uncohesive soils, such as sands and gravels, undisturbed samples are impracticable and the shear strength is normally assessed by use of the standard penetration test. In this method a borehole is sunk to a point about 450 mm short of the test level and a hardened steel sampler is then hammered down through the test stratum. The shear strength is measured by the number of blows (*N*) required to give a penetration of 300 mm, following 150 mm of seating drive.

Groundwater levels should be measured over a period, to allow for seasonal fluctuations, and this is done by inserting standpipes and either measuring by rod from the surface or, more conveniently, by use of an electronic dipmeter. A more sophisticated device is a piezometer, sealed into the soil with a bentonite plug, which enables readings to be taken of pore water pressures at various levels.

Groundwater level is not, of course, a sufficient indicator in itself of potential groundwater problems, and must be used in conjunction with information on soil porosity to assess the probable inflow of water into the trench excavation.

Deciding the intervals between site investigation borings is a genuine problem; they should be close enough to identify all significant changes and irregularities in the ground; but the cost of investigation can be substantial especially where the trench is very deep and perhaps tunnelling is an option. The cost-benefit of the investigation must be considered and there is no escape from the engineer's judgement in these matters. However, site investigation data are, more often than not, at the centre of contractual claims and disputes, so in cases of doubt the advice must be to err on the side of thoroughness.

4 Choice of excavation plant

The most significant change in pipelaying methods in the past 50 years has undoubtedly been the change from hand to machine excavation. Hand excavation is now only used in exceptional circumstances, such as where underground services are particularly congested.

There are two basic types of mechanical excavators — those with digging equipment operated from a jib and chain driven continuous trenching machines. In addition to these main types it is possible to excavate very narrow trenches using vacuum plant and in exceptionally wide trenches a further possibility may exist in that it may become economic to use scrapers, bulldozers and other earthmoving plant. However, this latter case will not be considered here in any detail as it is an exception to the general rule and the earthmoving project will have ceased to be a trench in the normally understood sense.

Continuous trenching machines

These are usually crawler mounted machines fitted with an arm which supports a system of continuously rotating chains to which brackets are fixed of appropriate width for the trench required (Fig. 4). The spoil is discharged to a windrow on one side of the trench, and because of the continuous nature of the cutting operation the spoil is reduced to fine tilth. This can greatly facilitate backfilling.

Jib operated excavators

In trenching terms jib operated machines may again be divided into two broad categories — draglines and backacters.

Draglines have a long jib controlled by cables which enables the bucket to be cast a considerable distance from the machine (Fig.

Fig. 4. Continuous trenching machine (courtesy New Civil Engineer)

5). This is the main asset of the dragline and it is used mainly in circumstances where length of reach is of paramount importance. Because the cutting edge of the bucket is powered only by the single directional pull of the lower ropes it does not cut hard material so effectively as the hydraulically powered backacter.

Backacters, sometimes called backhoes, consist of a power unit operating through a jointed arm which draws the bucket towards the machine (Fig. 6). Because of the configuration the excavated face is normally below the level of the machine, making backacters particularly well suited to trench excavation. The bucket itself can also be articulated relative to the jib making it possible to raise and deposit the contents into lorries or to spoil heaps above ground. Backacters can be either hydraulic or cable powered but hydraulic types are now almost universal.

Applications

The choice of machine for a particular project is influenced by the shape of the trench, battered or vertical, the type of ground, the type of pipes, and the jointing method to be used.

Fig. 5. NCK Rapier 406 crawler dragline discharging to tipper (courtesy Ransomes and Rapier Ltd)

Fig. 6. Backacter (courtesy Biggs Wall & Co. Ltd)

Continuous trenching machines can achieve high levels of productivity but require relatively soft homogeneous ground for maximum output. Hard obstacles can unbalance the machines and performance may be poor in coarse gravels or stony ground. Also these machines can only excavate, whereas draglines and backacters can use their jib to lift materials in and out of the trench. There are safety constraints on this lifting role as discussed in chapter 11 but there are many materials to be lifted, notably pipes, trench supports, and spoil for loading to lorries, so this can be a significant advantage. Where continuous trenchers are used, separate lifting equipment is needed, and the cost of this must be set against the extra speed of digging. However, trenching machines can be very effective on long cross-country pipelines where ground conditions are favourable. These include high-pressure gas and oil pipelines where the joints are butt-welded on the surface in advance of excavation. A group of machines then moves along the pipeline route, led by the continuous trenching machine. Bulldozers with side boom lifting arms then 'snake' the pipeline into the trench (Fig. 7). Behind these come a further bulldozer to backfill and compact the trench. High output can be achieved by this method but it is largely restricted to gas and oil lines. Water pipelines are seldom strong enough to resist the stresses involved in 'snaking in' and in any case are generally unsuitable for butt-welding, because of the need to ensure that the internal protection against corrosion is good.

The main application for draglines is where the trench has battered sides. In these conditions the long reach of the dragline and its ability to grade slopes below the machine level makes it the obvious choice.

For all the remaining types of trench, i.e. sewers and low and medium pressure gas and water mains, the backacter is the routine choice. These represent the overwhelming majority of all trenches and therefore the backacter may be regarded as the standard trench excavator. Backacters are robust machines, capable of operating in all kinds of ground except hard rock and are able to perform a number of useful extra functions, notably lifting and pushing home pipe joints providing that proper precautions are observed with regard to backfilling and partially compacting the trench.

Backacters are also relatively mobile. Traditionally they have

Fig. 7. Welded gas pipeline being laid by bulldozers with side booms (courtesy Laing Industrial Engineering & Construction Ltd)

been mounted on crawler tracks, but over the past 25 years there has been a trend towards wheeled types. These have the advantage of mobility and self-propulsion on public roads, and they can operate as effectively as crawlers in cross-country conditions. Rather surprisingly, however, their adoption in the UK has been slow compared with other European countries. For example, in

Table 1. Number of excavator sales for the period 1981-84

Type	UK	West Germany
Wheeled	545	6780
Crawler	7573	5690

West Germany, where operating conditions are similar, sales of wheeled excavators are much greater, as Table 1 shows.[7]

Subject to the limitations referred to, the machines described should be able to excavate trenches in most kinds of ground except hard rock. Fragmented rock can often be dug, with some difficulty, by a backacter bucket but more solid rock formations will need to be loosened before digging. This can be done using hand held pneumatic breakers, picks and wedges, or by a pneumatic breaker mounted on the bucket arm of an excavator. Where large quantities of hard rock are encountered, drilling and blasting may be necessary. The minimun size of explosive charge should be used and specialist advice should be taken to ensure that no damage or injury is caused to the public.

5 The causes of trench failure

Before the various options available for the support of excavated trenches are considered, the mechanisms by which trenches tend to fail, if unsupported, and the various forms such failures can take will be reviewed.

The most important factor to consider is the soil moisture of the adjacent ground and the changes which occur in the moisture content after excavation.[8] Soils consist of a mass of particles interspersed with voids (Fig. 8). In saturated conditions which, in northern Europe at least, are the most common, these voids are filled with water. If the ground is cohesive the voids are effectively sealed and if the soil is subject to a new, externally applied, load, the water pressure in the voids — the pore water pressure — will rise; but the water cannot escape and will continue to keep the soil particles apart. Therefore the friction between the particles is not increased and the shear strength of the soil remains unchanged. This is known as the undrained condition. However, if the soil is granular (i.e. sand or gravel) the water will escape under increased loading and consequently friction and shear strength will increase.

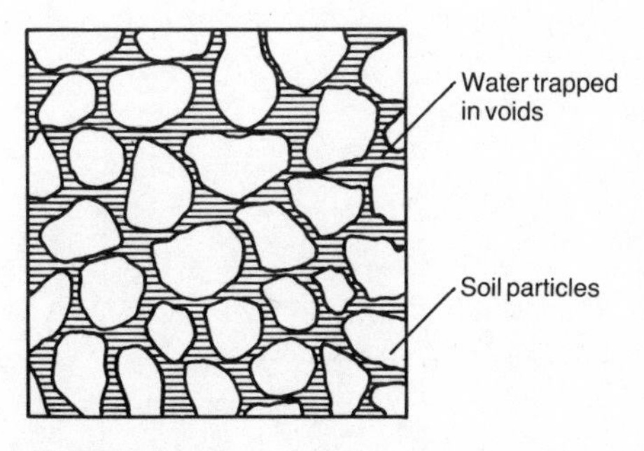

Fig. 8. Soil structure

This is the drained condition. The converse of the latter case is that if the external loading is reduced, the pore water pressure will decrease, additional water will be sucked in from adjoining ground, the friction between the particles will decrease, and so will the shear strength. This is what happens when a trench is excavated and the exposed trench side is, effectively, unloaded. Cohesive soils will maintain an undrained condition and remain stable, at least for a period, whereas granular soils will rapidly drain, lose shear strength, and collapse. The undrained condition will not, however, be maintained indefinitely, even in apparently impermeable clays. The draining process will proceed over a period of weeks or even months, depending on the soil permeability, and the trench will gradually weaken. It follows that the trench is at its most stable at the moment of excavation. In granular soils, for the reasons explained, failure will occur within a much shorter time and, at the extreme, fine sands and silts in saturated

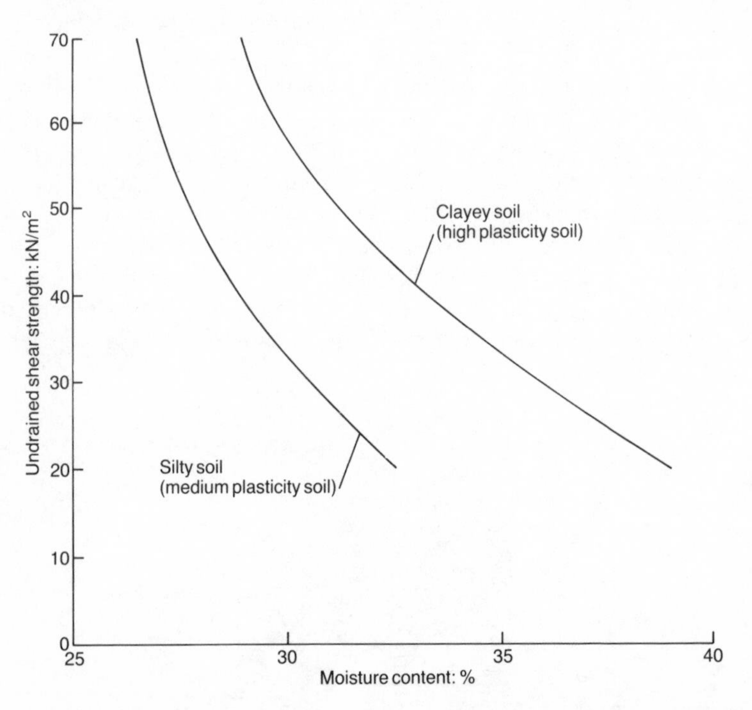

Fig. 9. Relationship between shear strength and soil moisture

conditions can become totally unstable when excavated giving conditions known as running sand.

The importance of the relationship between undrained shear strength and moisture content is illustrated by Fig. 9 which is taken from BS 6031.[9]

Some examples of how these processes work in producing failure of trench walls will now be considered.[10] In cohesive soils such as clays the vertical frame of a trench will tend, very gradually, to droop inwards (Fig. 10) under two influences: first under the effect of gravity as a plastic material under its own weight; and second through the swelling which occurs as moisture is sucked in from adjacent ground. Vertical fissures can develop from the ground surface accelerating the ingress of water. These conditions may eventually result in the sudden collapse of the trench wall, and this will occur more quickly if the ground has pre-

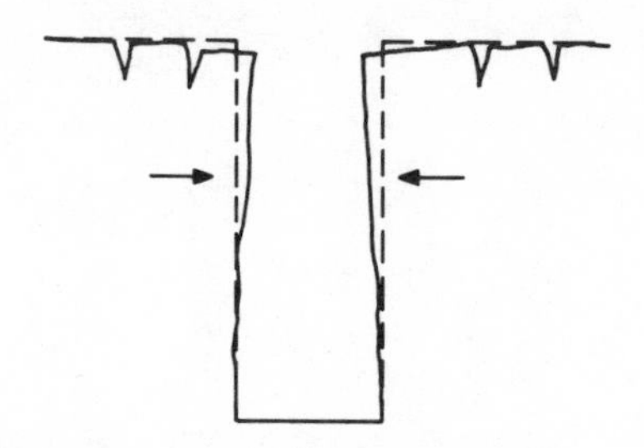

Fig. 10. 'Drooping' trench in cohesive clay

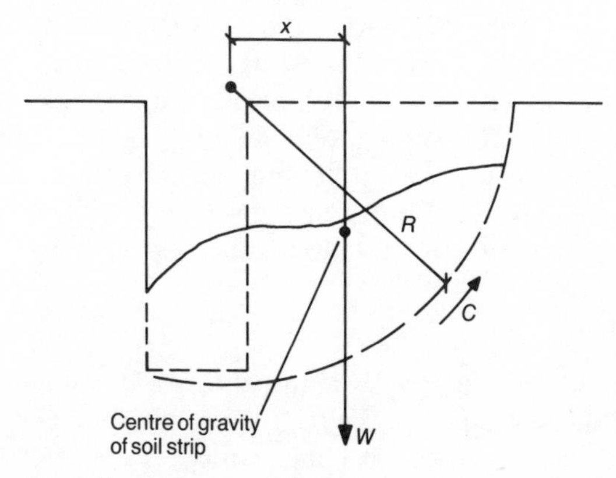

Fig. 11. Rotational slip

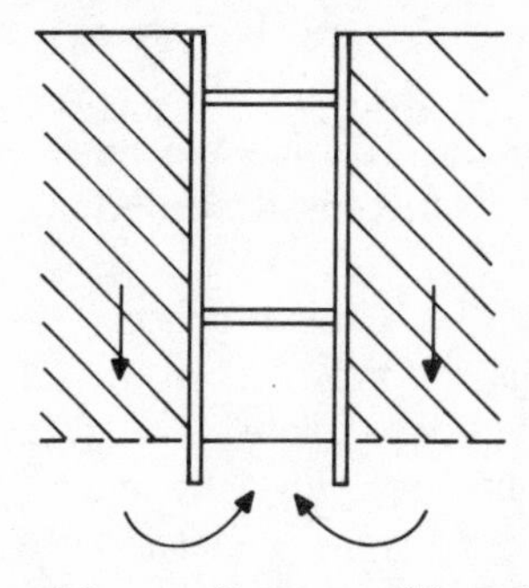

Fig. 12. Effect of load from adjacent soil in soft clays

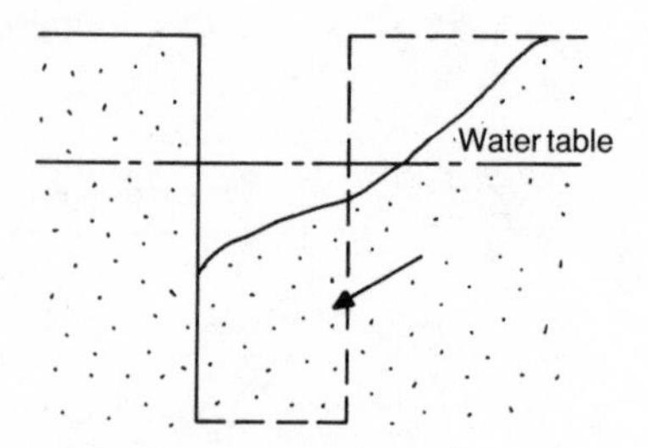

Fig. 13. Slump failure

existing fissures as a result of freezing and thawing cycles. In softer clays and silts failures can occur through rotational slips (Fig. 11) when the moment of the soil mass about the centre of rotation exceeds the restoring moment of the soil's shear strength. In soft clays there will be an upward pressure on the base of excavation produced by the weight of the soil block on either side which acts in the same way as the footings of buildings (Fig. 12). If the weight of these soil blocks is sufficient to overcome the bearing capacity of the soil at the trench bottom, heaving will occur.[11] The behaviour of a clay soil in these circumstances can be assessed by calculating its stability number N which is

$$N = \gamma H/C$$

where γ is the soil density, H is the effective depth and C is the undrained shear strength of the soil. When N approaches 6–7, the stability of the trench has become critical. In silty clays, or in mixed cohesive and granular soils with a high water table, the

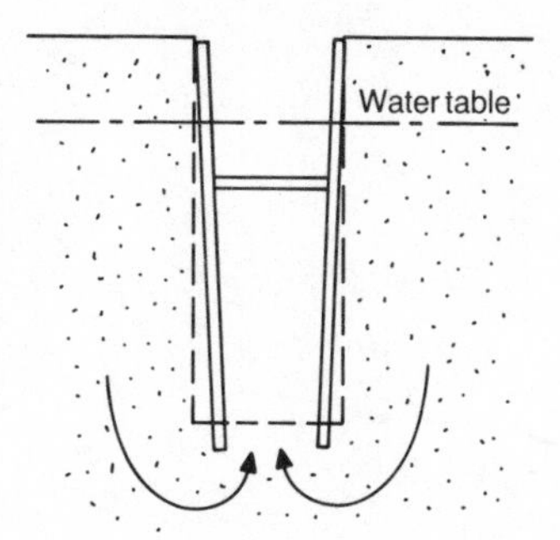

Fig. 14. 'Boiling' of trench bottom

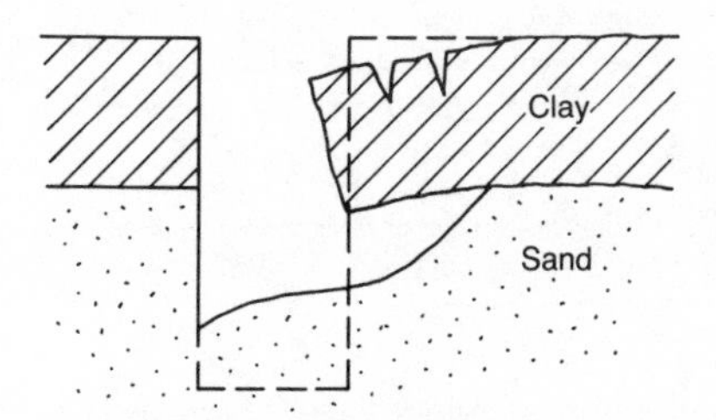

Fig. 15. Running sand condition

material can become unstable leading to slump failure (Fig. 13). In saturated sands and gravels, lowering the water table within the trench supports can lead to boiling of the trench bottom (Fig. 14) as the saturated soil moves under hydrostatic pressure. This condition can also arise, unexpectedly, when a thin permeable layer in the trench bottom is underlain by a porous layer subject to hydrostatic pressure.

When a layer of sand beneath the water table is exposed by the excavations the sand may slip into the trench bottom (Fig. 15) giving running sand conditions. This is probably the most difficult condition in which to excavate a trench.

It might be supposed that trenches excavated in rock would be stable and, indeed, they often are. However, failure can occur through bed irregularities, perhaps bands of weaker material (Fig. 16), or by displacement of blocks of rock along geological discontinuities such as bedding planes (Fig. 17). Soft rocks and chalk can weather and crumble if exposed for prolonged periods and clay or sand lenses in the trench wall can cause instability.

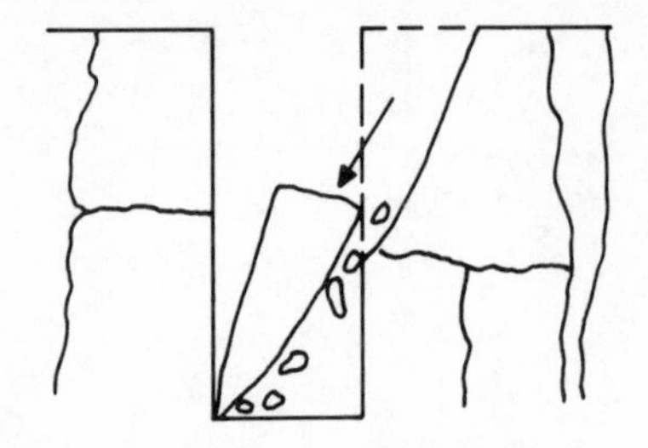

Fig. 16. Failure in rock through irregularities

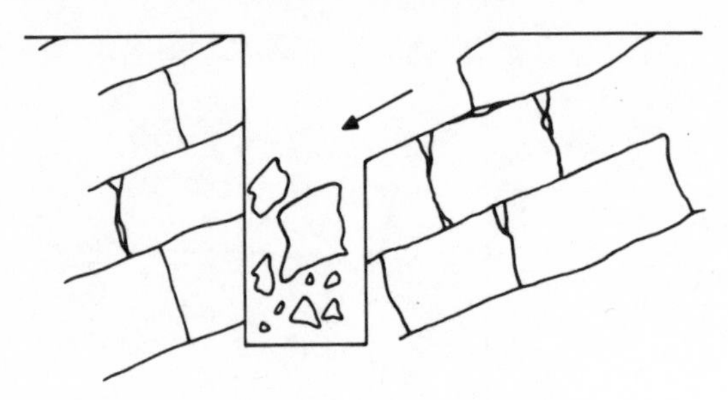

Fig. 17. Failure through slippage along bedding planes

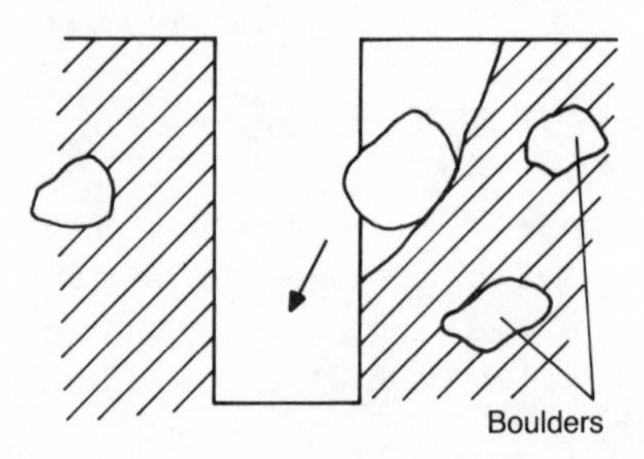

Fig. 18. Failure of trench wall through weight of boulders

Similarly, large boulders in otherwise cohesive soil can fall into the trench through their greater weight (Fig. 18).

In each of the conditions outlined above the probability of failure will be increased if the trench is dug along the contours of sloping ground (Fig. 19) or at the bottom of a slope, such as occurs in a highway cutting. The probability is also increased by superimposed loads on one of the trench sides. Such loads may arise from excavated spoil or construction vehicles being allowed to stand too close to the trench.

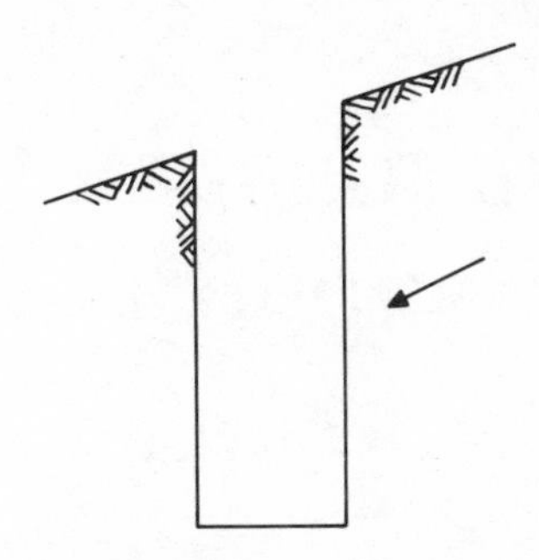

Fig. 19. Additional thrust on trench wall due to ground slope

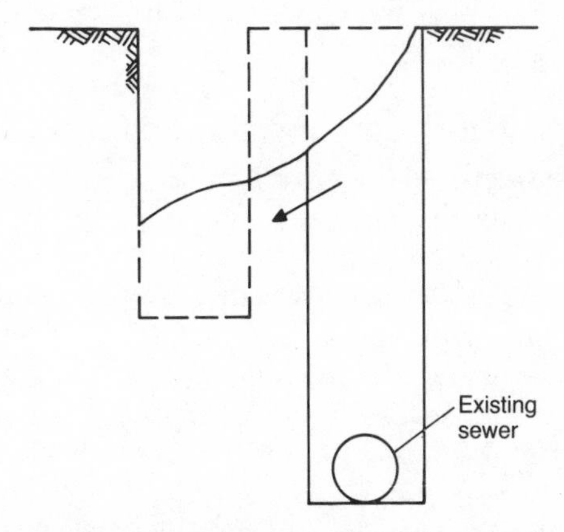

Fig. 20. Soft ground resulting from past excavations

Obviously the danger of collapse is also increased if the trench is situated, as many are, in a highway containing existing services. Even if these services were laid many years previously the soil above them is, effectively, made-up ground and will have less strength than virgin soil (Fig. 20).

6 Traditional methods of trench support

To prevent failures of the types described in chapter 5, trench support systems must be used and full advantage taken of the gratuitous benefit that trench walls can be propped against each other.[9]

Support by poling boards, waling and struts

Support systems now in use owe a good deal to former excavation practice. For many years the normal method of excavation was by hand and trench sides were supported by timbering, the components of which were poling boards, walings and struts as shown in Fig. 21. This method is still in common use except that the timber components have been largely replaced by steel, and the steel struts can be either mechanically or hydraulically expanded giving a more positive and reliable thrust on the waling.

This provides a convenient and versatile system of support. The components are relatively cheap and easily obtained. They are also light and can readily be moved within the trench. The system is flexible and can accommodate changes in the trench profile, local obstruction, crossing services, and constructional features, such as wider excavation for manholes in sewers.

However, the greatest advantage of this traditional method is that it can be taken down with the excavation, ensuring safe working at all stages. Of course, this advantage only applies where the excavation moves progressively downwards as happens with hand excavation. But hand excavation has now been almost eliminated in favour of backacters which advance the trench horizontally to its full depth, so that as the cutting line moves the trench sides are left, temporarily at least, unsupported. The insertion of conventional trench timbering, walings and struts can then be a dangerous operation unless special precautions are taken.

Such precautions could include the insertion of interim supports in the form of individual trench frames with hydraulically operated struts which can be fitted and expanded from the ground surface.

Another drawback of the conventional system is that its flexibility allows wide variations in design, i.e. in the arrangement and spacing of struts and walings on which decisions have necessarily to be made by very junior site supervisors.

A further constraint is the need to lower pipes, or other

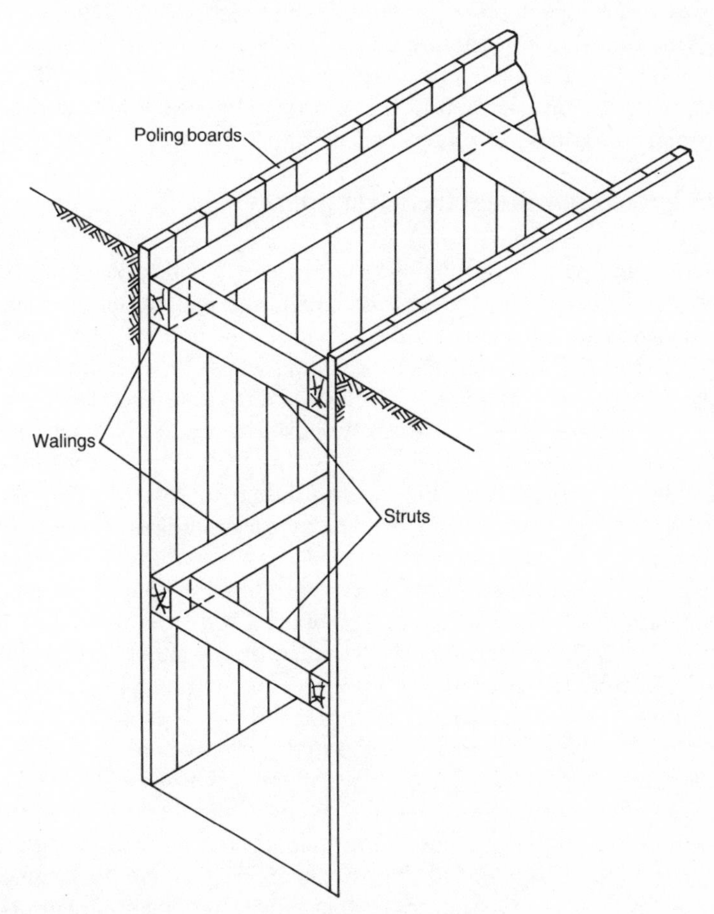

Fig. 21. Traditional trench support

equipment being installed, between the support struts. Traditionally, sewer pipes have been manufactured in relatively short lengths, normally less than 2 m, whereas cast iron pipes for water and gas are 4–6 m long, and steel pipes longer still. Obviously, trench supports could prove a serious obstacle to these longer pipes. Fortunately, however, most of them are laid as pressure mains and can therefore follow the ground contours in relatively shallow trenches with a minimum of side support. Deeper trenches with close strutting intervals are normally for sewers and the support system can be designed to pass the shorter gravity pipes. However, even gravity pipes are tending to increase in length and this may have its influence on future trends in support methods.

Despite these constraints traditional trench support using walings and struts remains the most common method and, if properly designed and placed, will give good service.

Design of traditional trench supports

The design of traditional trench supports cannot be precise because the pattern of loading from the supported soil can not be calculated accurately. It might seem logical to expect that the load would increase more or less directly in proportion to depth, as is assumed in the design of earth retaining walls. But site measurements have proved that assumption invalid for supported trenches.

Unlike retaining walls, trench supports are not a homogeneous structural unit but a complex series of individual supports which bear locally on the earth face with different intensities. The fact that the upper walings are inserted as excavation proceeds also means that the development of soil loading at these levels is effectively checked. At lower levels more soil movement occurs in the interval between excavation and bracing, whereas at the trench bottom support is continuously provided by the shear strength of the soil itself. Thus in stiff clay soils the maximum loading tends to occur not at the bottom, but at the mid-depth of the support system. Terzaghi and Peck[11] suggested envelopes of loading for clay soils (Fig. 22). These figures are acknowledged to be conservative. However, it would be possible to use them to calculate the initial pressures at any depth and hence the size of structural members needed to resist them. This can be done more simply by the use of design charts but, because of the great variety of soils and conditions which may be encountered, it should be

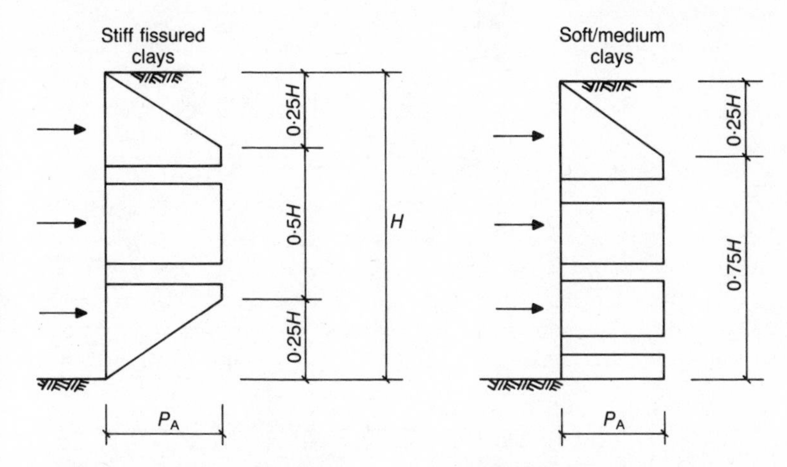

Maximum pressure $P_A = 0{\cdot}4\gamma H$ except where the construction period is short and the trench is well supported. In such cases P can be reduced to a minimum of $0{\cdot}2\gamma H$.

$P_B = K_A\gamma H$ where $K_A = 1 - m\,4c/\gamma H$ and m is a reduction factor dependent on the stress/strain characteristics of the clay; $m = 1{\cdot}0$ except where N exceeds about 4 under which conditions $m \leqslant 1{\cdot}0$

Fig. 22. Loading envelopes for clay soils

remembered that the design charts are based on assumed normal conditions such as a reasonably level ground surface, a limited surcharge load and an absence of water pressure behind the trench sheeting. It is important to check the design assumptions before using any such chart.

There are also deemed-to-satisfy design tables which are simpler to use but may not give the most economical solution. An example of these, taken from CIRIA report 97[10] is reproduced as Table 2.

In cohesive soils the pressure on the support system will tend to increase with time for the reasons given in chapter 5. If the trench is to be left open for a long period the increased pressure should be allowed for in the design. If a short-term trench support has to be left in place for unforeseen reasons, it should be inspected regularly and additional struts inserted if the walings show signs of excessive deflexion.

The pattern of walings and struts should be arranged in a balanced way to avoid any weakness or excessive pressure at the cantilevered ends (Fig. 23). To achieve this

Table 2. Reference table of walings and strut arrangements

(a) Unsaturated ground (except soft clays)

Maximum horizontal spacing of struts: m	Effective trench depth: m	Maximum vertical spacing of walings: m	Timber waling section: mm	Timber strut section: mm		
				Trench width up to 1 m	Trench width 1 – 1·5 m	Trench width 1·5 – 2 m
1·8	Up to 1·2	One set	225 × 75	150 × 75	150 × 100	150 × 150
			150 × 100	150 × 75	150 × 100	150 × 150
	3	1·0	225 × 75	150 × 75	150 × 100	150 × 150
		1·2	150 × 100	150 × 75	150 × 100	150 × 150
	4·5	1·0	225 × 75	150 × 100	150 × 150	150 × 150
		1·2	150 × 100	150 × 100	150 × 150	150 × 150
	6	0·9	225 × 75	150 × 150	150 × 150	150 × 150
		1·0	150 × 100	150 × 150	150 × 150	150 × 150
		1·2	250 × 100	150 × 150	150 × 150	150 × 150
2·5	Up to 1·2	One set	225 × 75	150 × 75	150 × 100	150 × 150
			150 × 100	150 × 75	150 × 100	150 × 150
	3	0·9	200 × 100	150 × 75	150 × 100	150 × 150
		1·3	Twin 225 × 75 spiked together	150 × 100	150 × 100	150 × 150
	4·5	0·9	Twin 225 × 75 spiked together	150 × 100	150 × 150	150 × 150

		1·5	225 × 150	150 × 150	150 × 150	150 × 150
	6	1·1	225 × 150	150 × 150	150 × 150	150 × 150
		1·5	300 × 150	200 × 150	200 × 150	200 × 150
3·0	Up to 1·2	One set	225 × 75	150 × 75	150 × 100	150 × 150
	3	0·9	Twin 225 × 75 spiked together	150 × 75	150 × 100	150 × 150
		1·5	225 × 150	150 × 150	150 × 150	150 × 150
	4·5	1·0	225 × 150	150 × 150	150 × 150	150 × 150
		1·3	300 × 150	150 × 150	150 × 150	150 × 150
	6	0·8	225 × 150	150 × 150	150 × 150	150 × 150
		1·0	300 × 150	150 × 150	150 × 150	150 × 150
		1·5	250 × 200	250 × 150	250 × 150	250 × 150
3·5	Up to 1·2	One set	200 × 100	150 × 75	150 × 100	150 × 150
	3	1·1	225 × 150	150 × 100	150 × 100	150 × 150
		1·5	300 × 150	150 × 150	150 ×150	150 × 150
	4·5	1·0	300 × 150	150 × 150	150 × 150	150 × 150
		1·5	250 × 200	200 × 150	200 × 150	200 × 150
	6	0·8	300 × 150	150 × 150	150 × 150	150 × 150
		1·1	250 × 200	200 × 150	200 × 150	200 × 150
		1·5	250 × 250	250 × 150	250 × 150	250 × 150

(b) Saturated ground (except soft clays and silts)

1·8	3	1·2	225 × 150	150 × 100	150 × 100	150 × 150
	4·5	1·2	250 × 200	150 × 100	150 × 150	200 × 150

(*a*) the first waling should be placed not lower than 500 mm below the ground surface
(*b*) the sheeting should be tied in at the base (this is normally done by pushing down with the excavator bucket)
(*c*) the waling should extend approximately 300 mm beyond the last strut to ensure that the strut load is evenly spread.

To assist installation of the walings, and to ensure that they remain at their correct level, steel hangers may be used, supported from higher walings or from the top of the sheeting. Alternatively, timber puncheons (vertical struts) may be inserted at intervals between and beneath the walings. The amount of positive support given to the trench side depends, of course, on how close a fit is achieved between the sheeting and the excavated face. The struts apply their pressure at intervals to the walings which are stiff members and may themselves be several metres long. Thus irregularities in the trench side may leave local areas unsupported, even though the struts are compressed. Timber wedges driven between the waling and the sheeting will help to ensure uniform

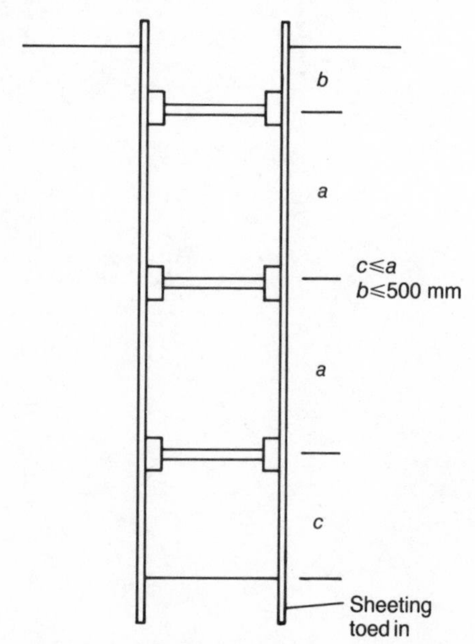

Fig. 23. Spacing of walings and struts

loading on the excavated face. Even if the sheeting is tightly wedged, however, it will seldom form a uniform junction with the soil face; but this does not mean that the support system is deficient. If a panel of trench sheeting were removed the exposed

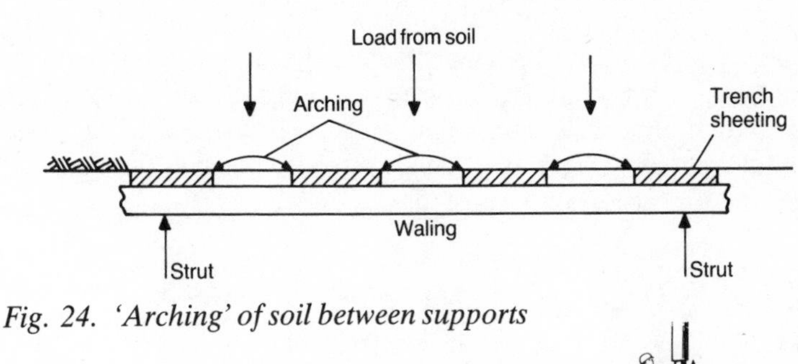

Fig. 24. 'Arching' of soil between supports

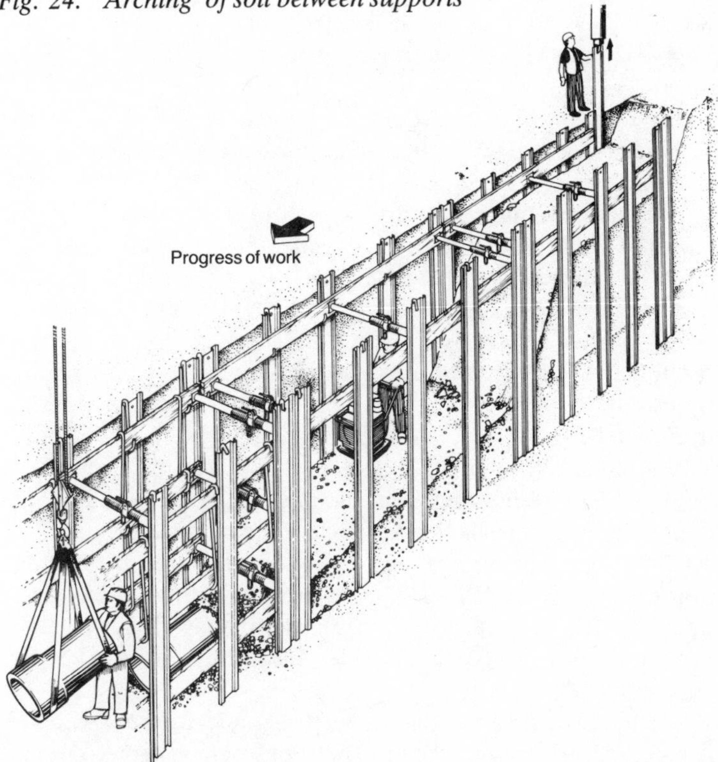

Fig. 25. Open sheeting support

section would probably remain in place for a considerable period. This is because the load previously resisted by the removed sheet has been transferred to those on either side by a process known as arching (Fig. 24). This principle is well demonstrated in firm soils or shallow trenches where substantial gaps are left between the vertical sheets. In its most common and traditional form this is known as hit and miss timbering or sheeting where alternate support panels are omitted but the walings and struts remain in place (Fig. 25). The interval between the sheet can be greater than this and alternatively individual strutted frames can be used without walings. Thus in stiff cohesive soils considerable areas of soil can safely be left unsupported.

Support by interlocking sheet piles

At the other extreme, uncohesive soils may not be able to stand unsupported even for the short period needed to insert the interim frames. The only way to excavate a trench in these circumstances is to drive sheet piling in advance of excavation. These piles will, of course, be a heavier section than normal trench sheeting and will require pile driving equipment to place. Interlocking steel piles are used enabling a watertight wall to be formed. Panels of 10–20 piles are driven within sets of guide timbers and the piles are pitched and driven two at a time. They can be driven to the full depth of the trench stage in one operation and walings and struts can be placed as excavation proceeds. Any soil or stones remaining in the troughs of the sheet piles after excavation should be removed, to prevent them being a hazard to operatives working in the trench. It is, of course, important that the piles are kept plumb during driving, both along the trench line and perpendicular to it; otherwise the trench profile may not be achieved and/or special taper piles may be needed to close the interlock.

As piles are stronger than the lighter section trench sheeting, they are less susceptible to damage and can, if required, be driven deeper than the normal toe-in depth of sheeting. This can be a decisive advantage where there is a tendency for the trench bottom to boil or heave.

Piling can provide a tight and effective support system with the soil/support load evenly spread. The disadvantages are considerably increased cost and a loss of flexibility in dealing with underground services. A further consideration is the noise of pile driving, which may be regarded as an environmental nuisance.

Soldier piles

A further variation, which provides a less cluttered working space within the trench, is the use of H-section soldier piles. These are pre-driven at intervals of about 2·5 m and the space between them is supported by horizontal sheeting wedged against the inner flanges of the H-section piles (Fig. 26). The piles themselves are steel universal beams between 200 × 200 mm and 300 × 300 mm in size and can either be driven to the depth required or can be lowered into an augered hole and concreted in below trench

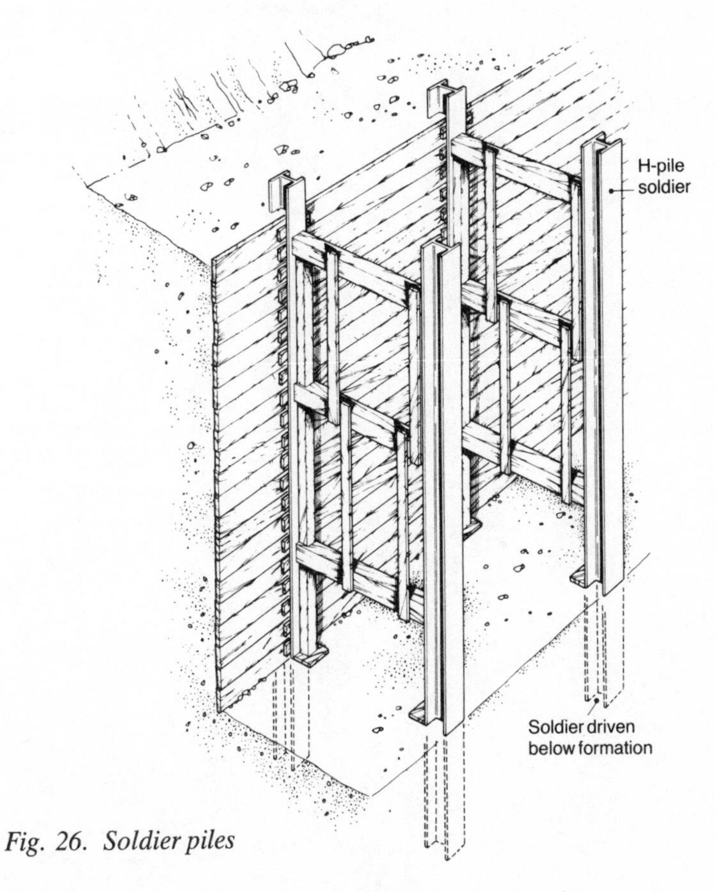

Fig. 26. Soldier piles

formation level. The latter method imposes a minor problem when the piles are withdrawn. The concreted section has to be cut off before backfilling or these sections of the pile must be pre-treated with a bond-breaking material. Excavation is taken down between the soldier piles and the horizontal sheets placed as excavation proceeds. It follows that, for this method to be successful, areas of soil below the last horizontal sheet must be capable of standing unsupported while the sheet beneath it is manoeuvred into position.

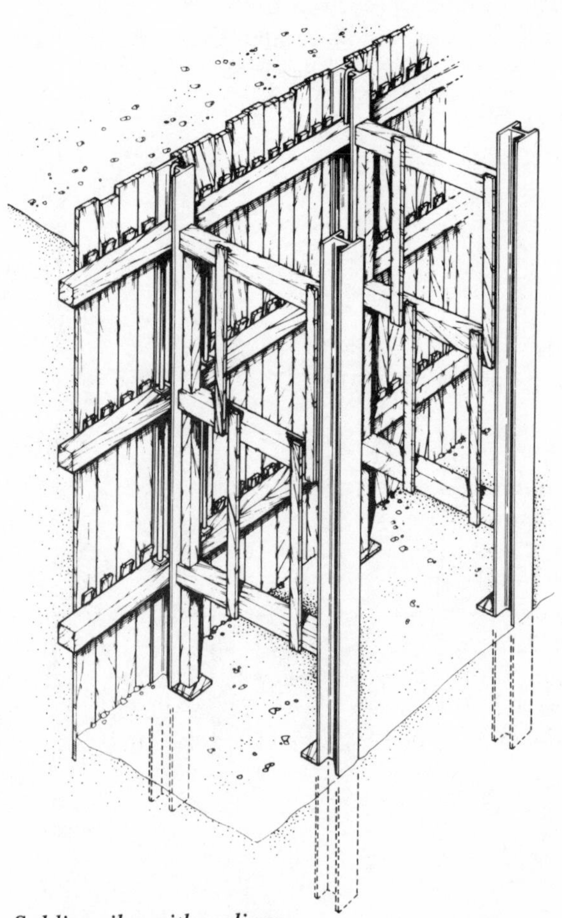

Fig. 27. Soldier piles with walings

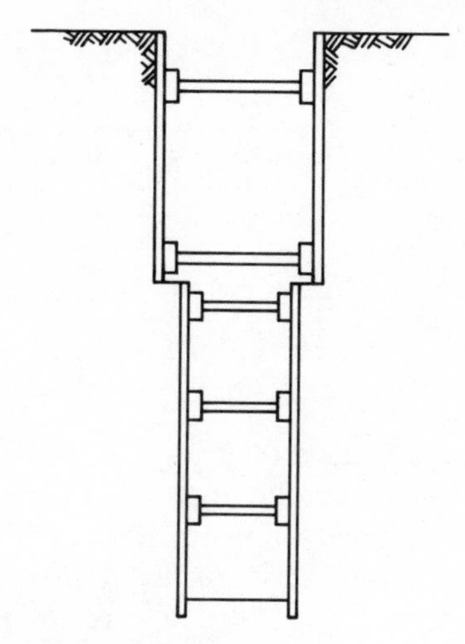

Fig. 28. Staged trench excavation

Alternatively, walings can be used spanning the flanges of the soldier piles and supporting vertical trench sheeting in the normal way (Fig. 27). The soldier piles are installed in pairs and braced by strutting between the pairs.

Staged trench excavation

All the support methods described so far comprise a single stage of excavation, i.e. a vertically sided trench taken down to the full depth required. In deeper trenches any of the methods described can be used in two or more stages (Fig. 28). The principles and methods of construction are the same in the lower stages but care must obviously be taken not to disturb the strutting of the upper stages and to allow space for excavation and for lowering pipes etc. between the more complex pattern of struts.

A simpler form of staging may be achieved by lowering the ground surface by a metre or so before starting to excavate the trench proper: an example of this can be seen in Fig. 6.

7 Proprietary support systems

The past two decades have seen a marked growth in innovative trench support systems which depart in varying degrees from the traditional methods discussed in chapter 6. Interest in these new proprietary systems has been fuelled partly by a commercial demand for labour saving construction methods and partly by a more conscious desire for site safety. The sytems available in 1979 are summarized in CIRIA report 95.[12] Proprietary systems can be grouped into three broad types

(*a*) hydraulically strutted shores and walings
(*b*) shields and boxes
(*c*) plate lining systems.

Hydraulically strutted shores and walings

Hydraulically strutted shores and walings are really a modest extension of traditional methods in that the struts and walings are pre-assembled in rectangular forms which can be lowered into the trench by the excavator between previously placed sheeting and the hydraulic struts are then expanded by remote operation from ground level (Fig. 29). This gives the considerable benefit that, provided that the sheeting can also be placed from ground level, operatives do not need to enter the trench during strutting. It is also claimed that only one banksman is needed for this operation. The same principle is used in vertical shoring frames for use where ground conditions do not demand continuous sheeting.

Manufacturers offer these frames in a variety of sizes up to 5 m long for walings and to over 2 m deep for vertical shores. The bearing faces of the vertical shores are also offered in a variety or widths, ranging up to 2 m. The hydraulic struts for all these systems can extend to support trenches up to more than 5 m wide.

Fig. 29. Hydraulically strutted shoring system (courtesy Mechplant Ltd)

Special frames are made to support larger excavations such as those for manhole construction. These often contain the hydraulic jacks within the telescopic members of the frame support, so as to leave a maximum clear opening for construction.

There are several such systems available in the UK and most of the components are made of aluminium alloy to save weight.

Shields and boxes

Shields and boxes are simple two sided rectangular structures placed in the trench to provide a safe enviroment for operatives to work (Fig. 30). Sometimes the walls can be jacked against the trench sides, but there may be a tendency for these to stick, particularly in deeper trenches, and where the support has been in place for several days. More often the box is designed simply as a safety device to protect operatives against trench collapse without any attempt to support the trench sides. There may be a significant gap between the trench wall and the base or the box may be stood clearly exposed in a battered wall trench. The loose fit makes it easier to drag the box along as excavation proceeds and this may offer a cheap and satisfactory system, provided that the excavator has the power to drag the box and that no damage is likely to result from the movement of the unsupported trench sides.

Fig. 30. Drag box (courtesy Mabey Hire Co. Ltd)

Several proprietary types of drag box are available, most having bolt-connected struts which can be adjusted to the trench width (Fig. 30). Some also have the front strut reinforced to provide a convenient hook-on point for the excavator bracket. The weight of the box can be important and this can be up to 3000 kg.

In addition to the manufactured products many contractors fabricate shields and boxes in their own workshops.

Plate lining systems

Plate lining systems consist of rectangular plate supports with one or two adjustable struts at each end (Fig. 31). The plates are supplied in vertical modules to accommodate a wide range of

Fig. 31. Plate lining system (Krings system)

trench depths and to enable the upper section of the trench to be supported as excavation proceeds.

In most cases the struts are designed to bear on reinforced sections of the plates themselves, but one manufacturer offers plates which slot into vertical H-section guide rails which are themselves strutted. These guide rails also serve to align the plates vertically. In other systems this is achieved by various forms of bolted connections.

Plate lining systems are available in a wide variety of unit sizes up to 5 m long and 2·5 m high, although most are considerably smaller. They are normally made of steel and can be extremely robust, capable of resisting external pressures of up to 40 kN/m^2.

A potential difficulty is that of strutting the bottom section of the trench while allowing space for the pipe to be installed. This is overcome in a number of ways. In some systems the lower modules have a bottom cutting edge which trims the trench wall as the excavator pushes it downwards and anchors itself securely in the trench bottom. An advantage claimed for this type is that the tapered lower section makes easy withdrawal possible if a concrete bed and haunch is placed.

In another system the upper support module is used, in effect, as a piling frame through which trenching sheeting is pushed and in a third system a concrete bed is recommended for larger pipes with the lower two modules tied with additional connecting bolts.

The above is a brief outline of some of the proprietary trench support systems now available. It is by no means an exhaustive summary as many variations exist, new models are constantly being developed and each manufacturer naturally claims advantages for his product. Some special systems have been produced which are much more sophisticated than those described and are capable of propelling themselves forward by jacking against previously backfilled sections.

If assembled by experienced operatives, proprietary systems may be quicker than traditional methods and because they can be operated more readily from the surface they should lead to safer working conditions. Against this, proprietary systems give less flexibility in dealing with existing services, withdrawal can sometimes prove difficult, and they are often unsuitable for use where ground movement adjacent to the trench is of particular concern.

8 Dealing with groundwater

As a pre-condition of orderly trenching operations, particularly those of preparing the trench bottom, placing and surrounding apparatus, it is clearly necessary to remove water from the excavation, or at least to reduce its quantity to a level which does not interfere with the construction. The trench support methods described in chapters 6 and 7 are based on the assumption that the water level has been reduced at least to the level of the trench bottom. Water can enter the trench

(*a*) from ingress of groundwater where the excavation penetrates the water table
(*b*) from surface water run-off
(*c*) from intersected land drains or ditches
(*d*) from flooding as a result of accidents, e.g. a burst water main
(*e*) from, in exceptional circumstances, excavation in impervious ground penetrating a previous stratum containing water under hydrostatic pressure.

Of these sources, the one requiring most serious attention is groundwater. This occurs to some extent in almost every excavation and the method of construction must be designed to deal with it.

At the site investigation stage information can be obtained on the standing water level in the ground and also on the permeability of the ground, which will affect the rate at which groundwater will flow into the excavation. It is essential to understand the distinction between these two properties.

The standing water level (or piezometric head) is the level to which water will rise in a borehole after allowing time to achieve equilibrium. This will vary with the season and possibly with adjacent river or sea levels. The flow of water into the excavation will depend on the dimensions of the trench and the permeability of the adjoining ground.

From the point of view of groundwater control the second property is the most relevant. A preliminary assessment of the flow can be made from the rate of change of water level in site investigation boreholes together with information obtained on the soil type. A cruder but sometimes more practical way of predicting groundwater movement is to excavate a series of trial holes or short trenches on the route of the permanent works and measure the rate of pumping required to reduce the water level below the proposed formation. Geological and other records can also be helpful, as can experience of other excavations in the locality. The problem of dealing with groundwater is reduced if excavation starts at the lowest point of the gradient and proceeds upwards.

In very permeable soils such as gravels and coarse sands the installation of interlocking trench sheeting or sheet piles will generally reduce the inflow. It is important that piles penetrate the trench bottom and they should preferably extend into a thick impermeable layer. If the penetration is insufficient the water flow will pass beneath the piles and result in boiling of the trench bottom as described in chapter 5.

For trench excavations in the normal range — up to 6 m deep — the methods used to control groundwater are[13]

(*a*) sump pumping
(*b*) well pointing.

Sump pumping

This is the simplest and cheapest method of removing groundwater and the one most commonly used. As the title implies the suction head of a pump hose is placed in a sump and the water is pumped to a disposal point at the surface.

Sump pumping can be used in all kinds of soil but in the case of water bearing gravel or coarse sand it will probably be necessary to install interlocking sheet piles to reduce the quantity reaching the pump. Sump pumping has the disadvantage that it draws water into the excavation and this can damage the formation and may leach out fine material from the adjoining ground. To prevent this the rate of flow can be reduced by installing additional sumps or by surrounding the suction hose inlet with a graded filter. The sump is normally formed near the side of the trench and usually consists of an open pipe surrounded with gravel. A piped drainage grip is sometimes constructed in the trench bottom to reduce the concentration of flow.

Well pointing

Well pointing is the method used when the groundwater flow is too great for sump pumping to work effectively. It consists of a series of tubes inserted vertically into the ground in a line parallel with the trench side and spaced at 0·6–2 m intervals. Each well-point unit comprises a metal tube about 40 mm dia. with a strainer section at the lower end. Each well-point riser is connected to a manifold to which suction is applied by a compressor pump (Fig. 32). Groundwater therefore flows by gravity to the well-point head and is sucked out through the manifold and discharged at ground level (Fig. 33). The suction head of the well point is below the trench formation level, and thus the excavation can be effectively drained.

In relatively soft soil the well points can be jetted into position using a high-pressure water pump. In harder ground a pre-bored hole may be necessary. The performance of the well point can be

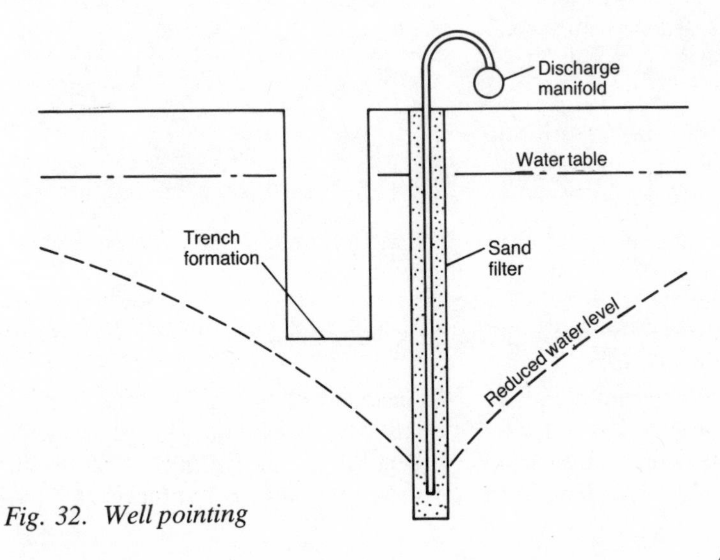

Fig. 32. Well pointing

Fig. 33. Example of well pointing (courtesy British Gas plc and the Water Authorities Association)

improved by surrounding the suction head with a screen of coarse washed sand.

Well pointing is normally installed on one side of the trench. Double sided systems can be used but these can seriously impede access to the trench side. Well points are at their most effective in sandy soils. Clays are unsuitable as the flow is insufficient for the system to operate properly, whereas in water bearing gravels the flow can be so great that the system becomes impracticable. The well points would have to be too close together to be economic.

Well pointing is a long established and well proved method of groundwater control and is normally operated by a specialist firm. It can effectively lower the water table by 4–6 m and has the advantage of providing a dry trench bottom. It is an expensive system to install and the operating cost is normally higher than for sump pumping because the quantity pumped is greater and no advantage is derived from any interlocking sheet pile barrier.

9 Bottoming-up and backfilling

Many failures of operational pipelines can be traced to poor preparation of the trench bed or to the pipe being surrounded by inappropriate material. Excessive bending moments can be induced in the pipe by isolated hardspots in the trench bottom or by point loads applied by hard material in the surrounding material. These stresses can be aggravated by traffic or other superimposed loading.

To avoid such problems the pipe should be placed on a bed which is as uniform as possible so that the load is evenly distributed. The material placed around and immediately above the pipe should be well compacted and free from hard elements such as large stones. Attention to these operations will go a long way to ensuring a long working life for the pipeline.

Where the soil is suitable, such as homogenous clays, the pipe may be bedded directly on the trench formation. The main excavation is taken down to within 50 or 100 mm of the formation and the remainder by hand. The level and gradient is set by sight rails erected across the trench with a traveller of appropriate length as shown in Fig. 34. Alternatively a laser unit can be set up within the previously laid pipeline and the line and level established by reference to a target placed inside the pipe being laid.

The uniformity of the final surface should be checked with a straight edge before the pipe is laid and joint holes of minimum length should be scooped out to ensure even bedding of the barrel.

Care should be taken to prevent the prepared formation becoming trampled or waterlogged before laying the pipe. Any soft spots should be removed and, in such cases or where the final surface has been overexcavated in error, the pipe should be adjusted to the correct gradient by inserting a layer of suitable granular material. Pipes should never be laid on bricks or other hard supports unless a concrete bedding is to be placed, as described later.

Where the ground cannot be trimmed to a uniform bed, the formation should be taken down sufficiently to enable a layer of bedding material to be placed to a minimum thickness of 100 mm beneath the pipe. This bedding material is normally either imported gravel aggregate or selected material taken from the excavation. Fig. 35 shows a typical pipe trench with layers separately identified for bedding and backfilling purposes.

The use of granular bedding material to support the pipe is a complex subject relating more to pipeline technology than to trench excavation. Discussion of trench backfilling could, however, be incomplete without a brief outline of the principles involved.[14]

Pipes used in buried pipelines can be divided into two categories, rigid and flexible. Rigid pipes, such as clay, concrete and cast iron are designed to carry all applied loading with a minimum of deflexion. Flexible pipes such as uPVC and polyethene are more susceptible to deformation and rely for their performance on the passive resistance of the material surrounding

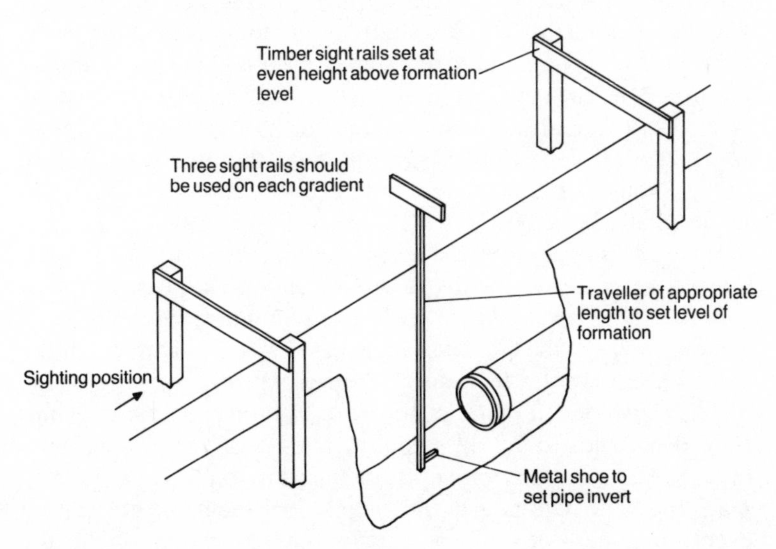

Fig. 34. Setting out trench gradients

them. As pipe diameters increase, the ratio of diameter to pipe wall thickness (the *d/t* ratio) also tends to increase so that the distinction between rigid and flexible pipes becomes blurred. Many large diameter pipes of rigid material, including ductile iron, require the support of selected sidefill. In fact, whatever the type or size of pipe, its load capacity will be increased in proportion to the degree of compaction achieved in the sidefill.

Effective compaction of the sidefill is therefore of great importance. The difficulty is, however, that the space available for equipment to operate between the pipe and the trench wall is severely limited. Furthermore if too much effort is applied the pipe can easily be damaged or its line or level disturbed. Therefore the material used for sidefill must be selected for properties which will achieve the maximum compaction under these conditions, i.e.

(*a*) it should be capable of being compacted around the pipe barrel with a minimum of applied effort
(*b*) the grading should be such that the fill material remains stable under loading from the pipe and when submerged beneath the water and that fines are not extracted by the passage of water.

In appropriate cases, selected as-dug material may be used for bedding and sidefill. This possibility is one to be considered at the time of site investigation and by monitoring the spoil during construction. Such material must be free of organic or combustible matter and should not contain large stones, large clay lumps or foreign soil. Compaction characteristics should be assessed by use of the compaction fraction test and the compaction fraction should not normally exceed 0·15.

Where it is necessary to use imported material this is normally gravel aggregate of suitable grading, but blast furnace clay or sintered pulverized fuel ash may be used for non-ferrous pipelines. The stone size of the fill material increases with the diameter of the pipe. Where heavier than normal pipes are to be supported, the bedding and sidefill material should contain angular or irregular shaped particles to ensure stability. These aggregates will, however, need more compactive effort than for rounded particles to reach the same density. The sidefill material should be brought up evenly on either side of the pipe, taking care not to disturb the line and level.

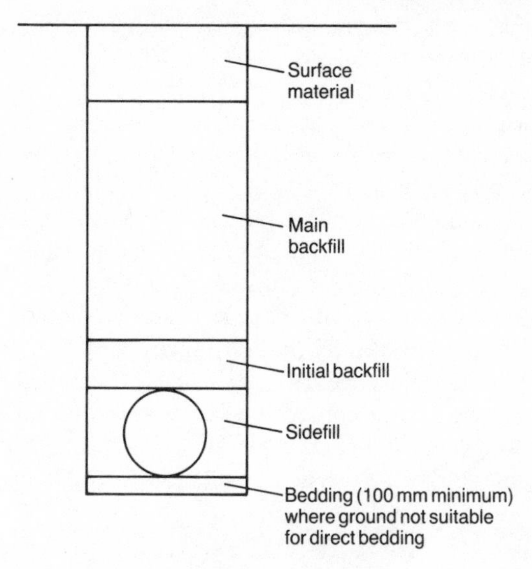

Fig. 35. Layers of trench backfill

The use of concrete as a bedding and sidefill material for pipes is not justified unless there are specific structural reasons, for example, the need to resist thrust from the foundations of adjoining buildings. When concrete is used the pipes should be placed to the correct line and level on concrete setting blocks immediately behind each socket with a layer of compressible material between the block and the pipe barrel. The blocks should be set so as to give a minimum of 100 mm depth of concrete beneath the pipe. It should be stressed that setting blocks are suitable only where concrete is to be used. In other circumstances they are bad practice and can lead to beam failure.

Where flexibly jointed pipes are used the flexibility should be maintained by placing layers of compressible material across the entire cross-section of the concrete to form flexible joints coincident with the pipe joints. The concrete should be of structural quality and should be placed in one operation, either as bedding, haunching, or complete surround as required.

In the layer of initial backfill, i.e. the layer 300 mm above the pipe crown, the material used should be similar to that used for sidefill and, in particular, should be free from large stones for the first 100 mm above the pipe. Compaction of the initial backfill should be carried out with light equipment such as hand tampers or light vibrators.

During all backfilling operations trench supports should be removed progressively as the fill material is placed and care should be taken to ensure that voids left by the supports are properly filled and compacted.

The remainder of the trench above the initial backfill layer should be backfilled and compacted in 300 mm layers up to the underside of the surface reinstatement layer.

Above the crown of the pipe concrete should only be used as backfill in exceptional circumstances and then only up to the minimum height necessary for structural stability. Backfilling the entire trench with concrete, even if lean-mix quality, is undesirable, particularly beneath carriageways, because the pipeline can be damaged by the direct transmission of traffic loads.

10 Reinstatement

The final act of a pipelaying project is the reinstatement of the ground to a condition in which it can be returned to its owners. This includes

(*a*) reinstatement of carriageways and footpath surfaces
(*b*) reinstatement of unpaved land, i.e. farm land, road verges and other open spaces
(*c*) reinstatement of land drainage.

Although reinstatement is a small element of the total project, the quality of reinstatement has a high public relations profile. If road or footpath reinstatement is poor, the public are directly affected and often have to endure the consequences for long periods. Relations with the farming community can also be damaged by poor reinstatement, particularly as it affects land drainage. Reinstatement work requires specialist skills such as placing of tarmacadam and agricultural cultivation which are scarce among pipelaying contractors.

It is therefore important to establish clearly what standard of work will be acceptable and it is a good idea, where possible, to arrange for the owners to do the finishing work themselves charging the cost to the promoter. Otherwise the employment of specialist contractors should be considered.

Highway reinstatement

Total success in reinstating road surfaces after trench excavation has proved an elusive ideal in the 90 years or so of metalled road surfaces. For a large part of that period subsidence of a backfilled trench was thought to be inevitable and the custom was to put in a temporary surface, consisting often of the excavated surface material covered with a thin tarmacadam topping. This was left to absorb traffic loads for many months, often up to a year or more, to allow settlement to finish. The surface was then broken out to sub-base level and replaced with new material by the highway authority.

This procedure, which is still sometimes followed, suffers from two major drawbacks. Because highway authorities are sceptical of the standard of compaction achieved, they tend to delay permanent reinstatement for a long period, sometimes well over a year, to be certain that settlement has finished. Road users thus have to suffer a substandard surface for an unnecessarily long period. More importantly, the tacit acceptance of a poorly consolidated trench also implies that the support given to the trench sides is weaker than that of the virgin ground. Hence ground movement towards the trench is likely to occur after backfilling, leading to longitudinal cracking in adjoining road surfaces. This does in fact occur all too frequently (Fig. 36) and few sights can be more dispiriting for a highway engineer. A further danger is that of damage to services in the adjoining ground which is described in chapter 11.

Fig. 36. Subsidence and road cracking adjacent to excavation (courtesy British Gas plc)

Since the mid-1970s there has been a concerted drive towards higher standards of compaction in highway trenches and the aim is to complete as much as possible of the permanent reinstatement as part of the construction process. A national specification has been agreed on these lines between representatives of the highway authority and the public utilities.[15] In areas where the model agreement has been signed, the trench is backfilled in layers to a good standard and the base courses of the highway surface replaced at the same time, leaving only the wearing course in a temporary condition to be subsequently replaced by the highway authority. In return the highway authority resumes responsibility for the trench surface six months after backfilling. Thus reinstatement should only be done once and done properly and the period of disruption of the surface should be reduced.

In a well ordered construction project it should be quite possible to achieve these standards. More difficulty will be found in emergency works such as the repair of burst water mains where the ground is saturated and the work has to be done under unfavourable conditions and time constraints. Particularly close supervision will be required in these conditions to meet the same standards.

To ensure that the reinstated tarmacadam surface is properly bonded to the existing layer, the surface should be trimmed back from the trench edge to expose a straight and vertical face. This exposed face is primed with hot bitumen before new courses of tarmacadam are placed, and the top courses should overlap those beneath by approximately 75 mm on each side (see Fig. 37).

In concrete carriageways the existing slab should be trimmed back to slope 10–30° to the vertical away from the trench to provide support for the reinstated area. Any reinforcement should be preserved, where possible, and linked into the reinstated slab.

Footpaths often have to be reinstated temporarily immediately after backfilling to maintain a safe passage for pedestrians. In the case of tarmacadam footpaths, the temporary surface can consist of a 40 mm layer of base coarse material, which can subsequently form part of the permanent reinstatement, if it remains in sound condition.

In concrete flagged footpaths the flags are set temporarily on granular material and subsequently lifted and re-set on lime mortar in the permanent position.

Reinstatement of unpaved land

Where the trench is in fields or open spaces a working strip is provided for the contractor. This represents the site and he has possession of it for a reasonable period during the contract.

In a narrower strip adjoining the trench the topsoil is removed and stored in a separate window and any good quality turf is lifted and stored for re-use.

On completion of pipelaying, the exposed ground beneath the topsoil layer will usually have been consolidated by the passage of construction traffic. This 'hard pan' will restrict the passage of groundwater and reduce the fertility of the ground. It should be loosened up by 'busting' with appropriate plant and it has been found that a proper shattering hoe operating at a depth of 450–500 mm gives better results than a ripper fixed to a crawler tractor.

Surplus spoil representing the volume of the pipe plus any imported material may have to be removed from the site. It is important to remember, however, that this material is the property of the land owner and may have a commercial value, e.g. sand. If spoil is to be disposed of, the wishes of the owner should be ascertained. Very often, however, it is possible to lose the

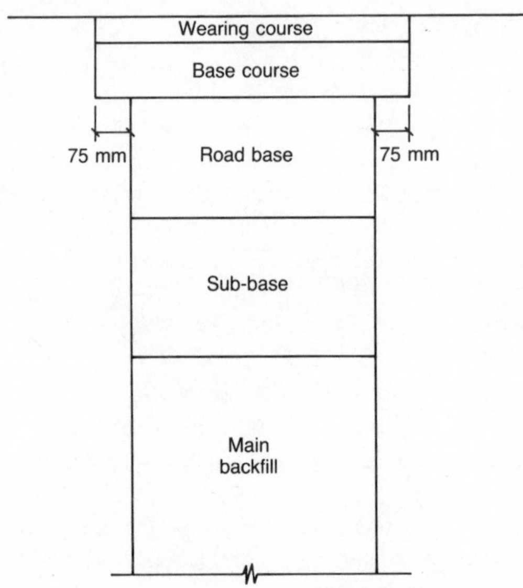

Fig. 37. Road surface reinstatement

surplus material by spreading it thinly over the stopped area before the replacement of topsoil.

The topsoil is cultivated, after replacement, to a standard acceptable to the engineer and the work of final cultivation, fertilizing and re-seeding, whether grass seed or other, is best done by the farmer himself.

Where turves have been removed the prospect of successful relaying depends on the time interval since cutting and on whether they have been watered during storage. In the winter months turf might survive two or three weeks storage but in the summer this might be reduced to one week.

Reinstatement of land drainage

Reinstatement of land drainage is one of the main areas of dispute with land owners arising from pipelaying. Not only can poor reinstatment cause problems at the time of construction but, with progressive settlement of trenches, flooding can develop several years later causing great difficulty in locating the fault and attributing responsibility. Land drainage is important to farming because excessive water in the soil can

(*a*) restrict aeration denying the soil micro-organisms and plant roots oxygen needed for respiration and induces accumulation of toxic substances
(*b*) retard the rise in soil temperature in spring, which is a critical growth stage, when all crops respond to warmer soil temperatures
(*c*) lower the load-bearing capacity of the soil. This can result either in cultivations being delayed or soil being damaged by cultivating under the wrong conditions
(*d*) aggravate nitrogen deficiency in crops
(*e*) damage seeds and seedlings by providing an environment conducive to attack by disease and pests
(*f*) limit depth of root penetration, thereby reducing the crops' resistance to drought.

In severe cases overland flooding can have a serious effect on crop yields. Flooding lasting longer than 24 hours can completely destroy a crop.

Trenching will inevitably sever existing land drains and additional damage can be caused to drains beneath land adjacent to the trench by the passage of heavy construction plant. Surface consolidation of these areas can also disturb the normal movement of air, water, and consequently soil nutrients.

At the planning stage it is important to discover as much information as possible about the drainage system and, where possible, to estimate the appropriate intervals at which land drains will be intercepted. Severed land drains can be reinstated by either reinstating individual drain pipes across the trench line, or by constructing an interceptor drain.

Reinstatement of individual drain pipes

This is the normal method of reinstatement and is appropriate where

(*a*) either there is no previous knowledge of the land drainage system or where it is known that the drains are fairly infrequent, say less that one every 40 m of pipeline trench
(*b*) there is little, if any, lateral movement of groundwater across the proposed trench line
(*c*) the reinstated trench will not form an impermeable barrier to the passage of water because of the soil type.

Before reinstatement, the trench should be backfilled and well consolidated up to the level of the severed drain. The ends of the existing drains should be exposed and cut back into firm ground until a section is exposed which is unaffected by the works. The replacement pipe is then placed so that it connects with the existing drain and bears on undisturbed ground for at least 500 mm at each end.

Pipes used for reinstatement should have the same diameter as the original drains and may be of any durable material. It is better, however, that they are long enough to span the trench without joints and pressure pipe off-cuts can often be used for this purpose.

Construction of an interceptor drain

In this method, no attempt is made to reinstate individual land drains. Instead a separate shallow trench is dug parallel to the main trench and on the upstream side of the intercepted land drains. The trench is dug to a depth of approximately 750 mm and is piped with open jointed land drains. It is then backfilled with single size stone ballast to within 350 mm or so of the ground surface (Fig. 38). This forms an interceptor or collector drain which picks up the flows from all existing drains and redirects it to discharge to an appropriate watercourse.

This method is appropriate where

(*a*) the number of land drains encountered or expected is large, (exceeding say one every 30–40 m).

(*b*) re-connection of the existing drain is difficult because the trench is exceptionally wide or is significantly deeper than average (i.e. more than 3 m) giving a greater risk of settlement; or the drains cross the trench at an acute angle; or are exceptionally deep and are intercepted by the pipe itself.

(*c*) soil conditions are such that the reinstated trench will present a barrier to lateral ground movement.

It will often be appropriate to lay sections of interceptor drain selectively along parts of the pipeline route while reinstating individual drains for the remainder. The work should be done by a specialist contractor who may be expected to construct the drain at the rate of approximately 200–300 m per day. Higher rates can be achieved under ideal conditions.

There is room for debate as to whether the interceptor drain should be laid before or after pipelaying. The normal sequence is after pipelaying which has the advantage that the full extent of the severed existing drainage is known and that subsequent damage to the drain and to any associated mole drainage can be avoided.

Construction of the drain before pipelaying is less common but

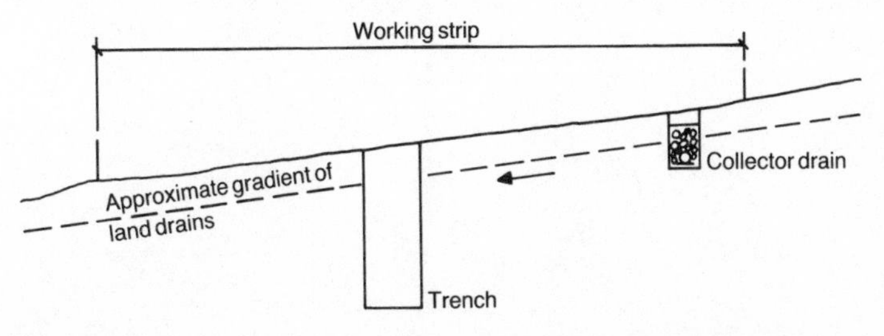

Fig. 38. Use of interceptor drain to reinstate land drainage

can be appropriate in certain circumstances, for example, where it is known with some precision that a large number of existing drains will be intercepted or where the groundwater flow is exceptionally large, e.g. from upstream springs. The technique has a further advantage in that the pipelaying trench itself is protected from ingress of land drainage water thus avoiding the need for temporary bridging connections. A drawback is that the engineers' freedom to amend the pipeline route for any unforeseen reason is restricted. Also interceptors laid before pipelaying must be placed as close to the edge of the working strip as possible whereas those laid afterwards can be closer to the trench. This latter point can be significant in the case of exceptionally wide working strips, as the catchment between interceptors and the pipe trench may be large enough to produce its own drainage problems.

Reinstatement of land drains can be accomplished satisfactorily by any of the methods discussed above. The traditional method of reconnecting individual drains continues to be the economic solution in most circumstances and requests by owners and agents for interceptor drains should only be accepted where the criteria outlined are met. However, it is also true that interceptors are relatively cheap to install and normally offer a permanent solution which is recognized and appreciated by the farming community. Their public relations value should therefore be borne in mind.

Above all it is vital that a clear decision should be made at the planning stage as to what sort of reinstatment is to be used. If the decision is deferred until the trench is excavated there is a risk that the pipeline promoter will end up with the worst of all worlds, having paid first for reconnection, then for an interceptor as remedial work and finally for loss of crop due to the occurrence of flooding while the matter is being resolved.

11 Contractual aspects and safety

As mentioned in chapter 1, trench excavations unfortunately contribute regularly to accident statistics as Table 3 illustrates. The achievement of consistently safe working conditions should therefore be a primary aim of all concerned.

The contractual relationship

Under all forms of contract, responsibility for site safety lies squarely with the contractor. This must necessarily be so because the contractor employs the site staff and gives them their instructions. Furthermore, although temporary works, choice of plant and method of working are subject to the approval of the engineer, they are matters for the contractor's expertise and to be decided primarily by him. Approval by the engineer does not relieve the contractor of responsibility for safe working or indeed for the effectiveness of working methods in other respects.

Regrettably, as in other areas of work attention to safety tends to delay progress and can be costly. With construction capacity often exceeding demand there is pressure to reduce contract prices and ultimately to cut corners.

Table 3. Fatal accidents in the UK from trench collapse

Year	Number of fatal accidents
1981	5
1982	4
1983	8
1984	8
1985	6

Because of this tendency the employer and the engineer also have responsibilities when awarding and supervising the contract to ensure that the work is let to a reputable firm, that the contractor's proposals, submitted for approval, take proper account of safety hazards, and that during construction safe practices are, in fact, followed. This latter duty can be a testing one for supervisory staff as it can involve grey areas and matters of opinion as to what is, or is not, safe. An over-zealous supervisor can find himself faced with claims if the contractor thinks he has overstepped the boundary of what is reasonable and necessary. There is, however, no escape from these decisions by resident site staff and in the event of an accident they will certainly be called to take their share of blame if this duty has been neglected.

Safety hazards in trench excavation

Hazards to men working in the trench

A trench is potentially a dangerous environment in which to work (Fig. 39) as accident statistics show. Trench walls can collapse with little warning and a man buried by such a collapse has only a short time, perhaps only minutes, in which to be rescued. Such accidents can be largely eliminated by trench support, provided that the support system is adequately designed and does not require men to enter the trench to insert the initial support frames. Other precautions include the avoidance of superficial load from construction vehicles (Fig. 40), excavated spoil or stored pipes being allowed to bear on the ground surface too close to the trench. These superficial loads should be kept at least 1·5 m from the trench side.

Falling objects are another hazard to trench operatives. Pipes, trench support components and other equipment such as pumps have to be lifted in and out of the trench while work is in progress. Suitable lifting plant and trained operatives are therefore essential for safe working. Separate lifting plant, in addition to excavators, can be expensive and there is a strong commercial case for using the excavator jib as a crane substitute. This can be done safely provided that the machine is operating within its design working load and provided that a Certificate of Exemption has been issued under the Construction (Lifting Operations) Regulations 1981–82. For compliance with the terms of this certificate hydraulically operated machines must be fitted with check valves to prevent a gravity fall in case of hydraulic failure. Plant without such a certificate should not be used for lifting. As a further precaution the whole of the site, including the trench, should be a hard hat area.

Fig. 39. Dangerous unsupported trench (courtesy British Gas plc and the Water Authorities Association)

Hazards from existing underground services

The underground apparatus of most public utilities are live systems and the principal services, electricity, water and gas can be dangerous to trench operatives when damaged or fractured. All the utilities keep records of their assets but these are of variable quality and in some cases are not accurate enough to be relied on. Ferrous mains and power cables can be detected from the surface

Fig. 40. Trench at risk from vehicle load (courtesy British Gas plc)

by electrolocation equipment, but for others the only certain location method is by pilot trench holes ahead of main excavations.

Fig. 41. Gas main holding back trench wall (courtesy British Gas plc and the Water Authorities Association)

In the case of electricity cables the dangers of violent contact are obvious but, fortunately, given reasonable care, they can usually be located clearly enough to avoid damage.

Gas and water mains present more complex problems and for them location and avoidance may not be sufficient precautions. Research shows that, even with the most effective system of trench support, some degree of movement in the adjoining ground may be the inevitable result of deep trench excavation. In an urban

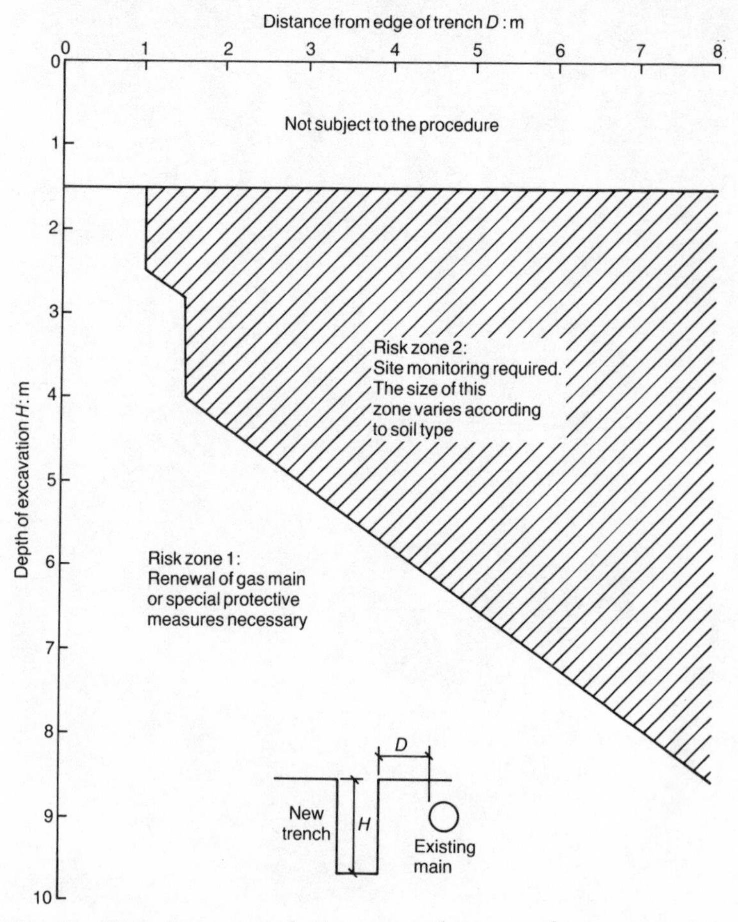

Fig. 42. Water/gas consultative procedure on deep excavations (courtesy British Gas plc and the Water Authorities Association)

environment the ground will probably contain gas and water mains of grey cast iron, which is an intrinsically brittle material. Ground movement resulting from excavation may be sufficient to cause these mains to fracture, especially if they are old and suffering from internal or external corrosion. Misuse by a contractor can be a further hazard (Fig. 41).

Fracture of a water main under normal pressure can lead to sudden inundation of the trench and adequate means of egress should be provided to enable people in the trench to escape quickly. Fracture of a gas main can be even more serious. An investigation into a number of serious gas explosions in the late 1970s concluded that some of these were caused by deep excavation in adjoining ground. Most such deep excavations are for sewers.

Many of these fractures and explosions occur not only during the trench excavation but also a considerable time afterwards as the adjacent soil adjusts to the environment of the reinstated trench. Arrangements aimed at avoiding such damage have been agreed nationally between the gas and water industries and are formalized in a consultative procedure[16] which provides for various precautions including replacement of mains which lie within a certain range of a proposed deep trench (Fig. 42)

When existing services cross a trench line, or are otherwise undermined by the excavation, they should be supported either by propping from beneath or, where this is impracticable, by hangers suspended from a beam at the ground surface. Temporary support of this kind is the contractor's responsibility, but if the position of the existing service is such as to require some form of permanent support this would normally be the employer's responsibility and should be designed by the engineer. In all such cases the agreement of the utility in question should, of course, be obtained.

Hazards to existing buildings

Any deep excavation near an existing building brings with it the risk of undermining the foundations, leading to settlement damage and, in extreme cases, to the collapse of the building. Obviously the closer the excavation to the building, the greater the risk, and within a certain range special precautions must be taken. The normal procedure is to assume that the load from the foundation passes downwards through a wedge of existing ground and that

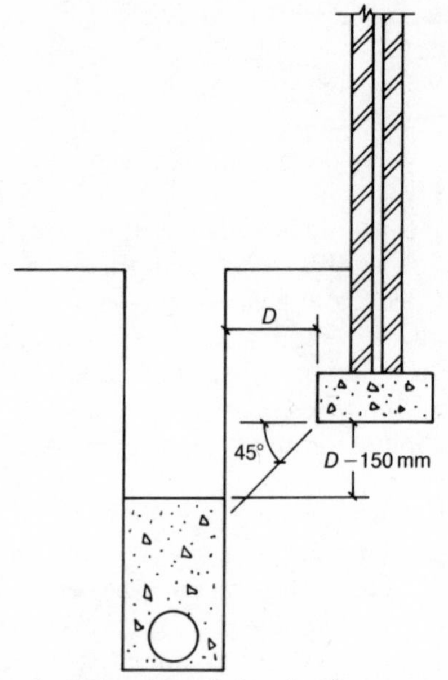

Fig. 43. Effect on building foundation loads on trench

any penetration of this wedge by the trench excavation represents a hazard.

The normal remedy when this occurs is to backfill the trench with concrete up to the level at which the perimeter line of the supporting wedge cuts the trench. A slope of 45° has often been assumed, but a recent British Standard[17] takes a slightly more conservative view and recommends an additional 150 mm of concrete above the 45° line (Fig. 43).

Hazards to the public at the surface

Trench excavations can represent a danger to members of the public in their normal use of the ground surface. In highways they can be a danger to vehicles unless they are properly marked and illuminated. When the excavation occupies a substantial part of the road temporary traffic lights may be needed or even closure of the entire road. These arrangements require prior consultation and agreement with the highway authority and the police. The scale and arrangement of warning signs for road works is laid down in the *Traffic signs manual*[18] a typical extract of which is shown in

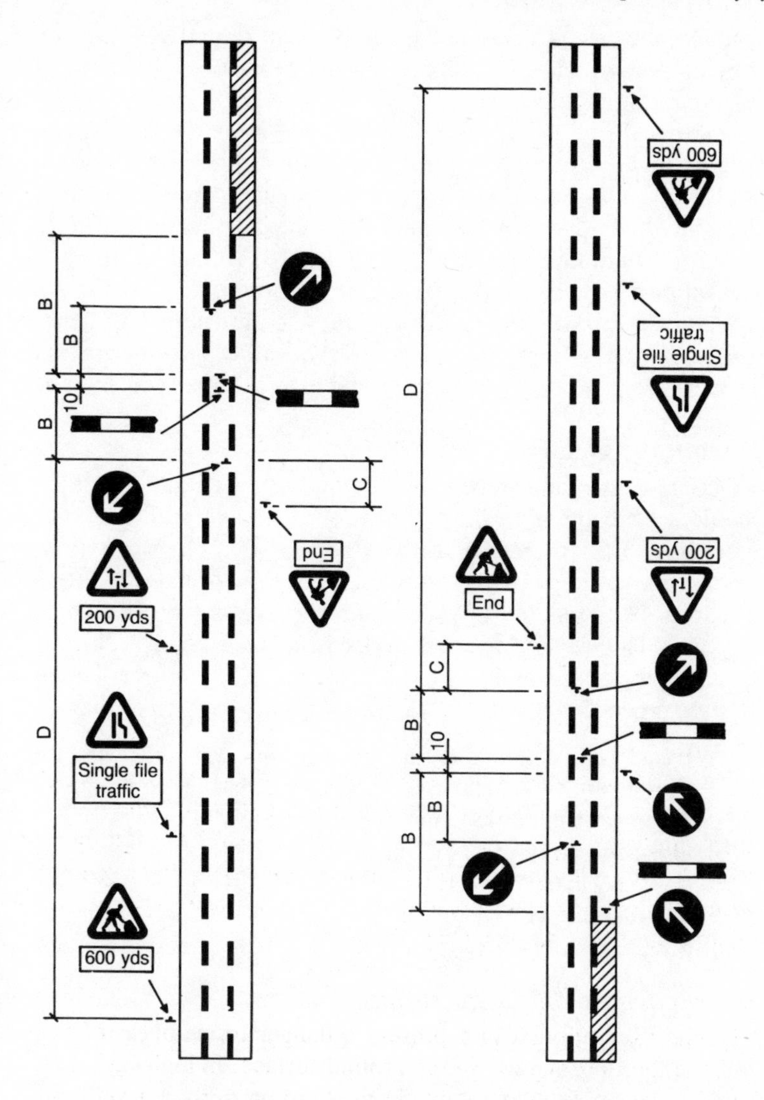

Fig. 44. Typical extract from chapter 8 of reference 18: layout of signs for road works on single three-lane carriageway road, one lane closed at edge of road; lettered dimensions are tabulated in reference 18 for various conditions (courtesy Controller, HMSO)

Fig. 44. However, it should be noted that the *Traffic signs manual* is being revised and the extract reproduced here does not reflect current requirements.

Guard rail protection should be installed around open excavations in highways, public open spaces and all areas to which the public or pedestrians have access. These guard rails can often conveniently be fixed to the trench supports and should extend to at least 1 m above normal ground level. At night, an open excavation should be lamped for the same purpose.

Where the trench passes through private agricultural or other land the form of temporary fencing should be agreed with the owner or occupier of the land according to local circumstances.

The human element

Decisions affecting trench support and other detailed working practices have often to be taken at short notice and by junior operatives. It is therefore most important that they and their supervisors are adequately trained and understand the basic principles of the support system and of other equipment which they use. The site foreman or supervisor has a key role in this respect. He should make regular inspections to ensure that no slipshod practices are developing and to check in particular that

(*a*) trench support systems are performing correctly with no sign of undue strain or deflexion, and that all components are in good condition and properly aligned
(*b*) all wedges are tightly driven
(*c*) no heavy load is placed near the trench edge
(*d*) crossing services are properly supported
(*e*) access into and across the trench is being properly maintained
(*f*) hard hats are being worn
(*g*) the excavation is properly fenced and signed by day and, in addition, adequately lit at night.

Compliance with construction regulations

Because of the seriousness of the risks involved, some of the precautions outlined are made compulsory by the Construction

Regulations 1961 which have statutory force. Briefly summarized, these require that

(*a*) trenches deeper than 1·2 m should be supported (with qualified exceptions for battered sided trenches and for operatives actually placing the supports)
(*b*) supports should be installed under the supervision of a 'competent person' who should also inspect the excavations regularly.
(*c*) means of rapid exit from the trench should be provided where a sudden rise in water level is possible
(*d*) precautions should be taken against damage to adjacent buildings
(*e*) excavations should be fenced in some circumstances
(*f*) no material, equipment or load should be placed near the edge of the excavation.

Acknowledgements

The author is grateful to Peter Rumsey of the Water Research Centre, Swindon for reading the manuscript and making helpful suggestions.

The following organizations have granted permission for reproduction of photographs, sketches and tables and their assistance is acknowledged with thanks

Biggs Wall & Co Ltd
British Gas plc
British Standards Institution
The Construction Industry Research and Information Association
Krings Verbau
Her Majesty's Stationery Office
Laing Industrial Engineering & Construction Ltd
Mabey Plant Hire Co Ltd
New Civil Engineer
Pipe Jacking Associaton
Ransomes and Rapier Ltd

Figures 9 and 21 have been reproduced by permission of the British Standards Institution, 2 Park Street, London W1A 2BS, from whom copies of the British Standard BS 6031 may be obtained.

Figure 3 has been reproduced by permission of the Pipe Jacking Association, Suite 411, London International Press Centre, Shoe Lane, London EC4A 3JB.

Table 2, Fig. 25 and Figs 26 and 27 have been reproduced from pages 31, 33 and 43, respectively, of CIRA report 97 *Trenching practice* by permission of the Director General of the Construction Industry Research and Information Association.

References

1. Fry R. and Rumsey P. B. *Ground movement caused by trench excavation and the effect on adjacent buried services.* Water Research Centre, Marlow, 1984, WRC TR 1950.
2. Glennie E. B. and Reed K. Social costs: trenchless v. trenching. *No-dig 87*. Water Research Centre, Marlow, 1985.
3. Robb A. D. *Site investigation*. Thomas Telford, London, 1982.
4. British Standards Institution. *Code of practice for site investigation.* BSI, London, 1981, BS 5930.
5. Dumbleton M. J. and West G. *Preliminary sources of information for site investigation in Britain.* Transport and Road Research Laboratory, Crowthorne, 1976, report 403.
6. British Standards Institution. *Methods of test for soil for civil engineering purposes*. BSI, London, 1975, BS 1377.
7. Review of excavation plants. *Contract J.*, 1986, 19 Sept.
8. Rumsey P. *Soil behaviour and trench support*. Water Research Centre, Marlow, 1982, WRC ER 70E.
9. British Standards Institution. *Earthworks*. BSI, London, 1981, BS 6031.
10. Irvine D. J. and Smith R. J. H. *Trenching practice*. Construction Industry Research and Information Association, London, 1983, report 97.
11. Terzaghi K. and Peck R. B. *Soil mechanics in engineering practice*. John Wiley and Sons, New York, 1967.
12. Mackay E.B. *Proprietary trench support systems*. Construction Industry Research and Information Association, London, 1982, report 95.
13. Somerville S. H. *Control of groundwater for temporary works*. Construction Industry Research and Information Association, London, 1986.

14. Water Authorities Association. *Imported granular and selected as-dug bedding and sidefill materials for buried pipelines*. WAA, London, 1987, I&G note 4-08-01, issue 2.
15. Department of the Environment. *Reinstatements under the Public Utilities Street Works Act 1950: a model agreement and specification*. HMSO, London, 1974.
16. Water Authorities Association/British Gas Corporation. *Model consultative procedure for pipeline construction involving deep excavations — second revision*. WAA/BGC, London, 1985.
17. British Standards Institution. *Building drainage*. BSI, London, 1985, BS 8301.
18. Department of the Environment/Scottish Development Agency/Welsh Office. *Traffic signs manual*. HMSO, London, 1973, ch. 8.
19. Water Authorities Association. *Civil engineering specification for the water industry*. WAA, London, 1981, 2nd edn.